# Lecture Notes in Computer Science 4771

*Commenced Publication in 1973*
Founding and Former Series Editors:
Gerhard Goos, Juris Hartmanis, and Jan van Leeuwen

T0223119

Thomas Bartz-Beielstein
María José Blesa Aguilera
Christian Blum   Boris Naujoks
Andrea Roli   Günter Rudolph
Michael Sampels (Eds.)

# Hybrid Metaheuristics

4th International Workshop, HM 2007
Dortmund, Germany, October 8-9, 2007
Proceedings

 Springer

Volume Editors

Thomas Bartz-Beielstein
Fachhochschule Köln, Germany, E-mail: Thomas.Bartz-Beielstein@fh-koeln.de

María José Blesa Aguilera
Universitat Politècnica de Catalunya, Spain, E-mail: mjblesa@lsi.upc.edu

Christian Blum
Universitat Politècnica de Catalunya, Spain, E-mail: cblum@lsi.upc.edu

Boris Naujoks
University of Dortmund, Germany, E-mail: boris.naujoks@uni-dortmund.de

Andrea Roli
University of Bologna, Italy, E-mail: andrea.roli@unibo.it

Günter Rudolph
Universität Dortmund, Germany, E-mail: guenter.rudolph@uni-dortmund.de

Michael Sampels
Université Libre de Bruxelles, Belgium, E-mail: msampels@ulb.ac.be

Library of Congress Control Number: 2007936297

CR Subject Classification (1998): F.2, F.1, G.1.6, G.1.2, G.2.1, I.2

LNCS Sublibrary: SL 1 – Theoretical Computer Science and General Issues

ISSN        0302-9743
ISBN-10     3-540-75513-6 Springer Berlin Heidelberg New York
ISBN-13     978-3-540-75513-5 Springer Berlin Heidelberg New York

Springer is a part of Springer Science+Business Media

springer.com

© Springer-Verlag Berlin Heidelberg 2007
Printed in Germany

Typesetting: Camera-ready by author, data conversion by Scientific Publishing Services, Chennai, India
Printed on acid-free paper      SPIN: 12170852      06/3180      5 4 3 2 1 0

# Preface

The International Workshop on Hybrid Metaheuristics is now an established event and reaches its fourth edition with HM 2007. One of the main motivations for initiating it was the need for a forum to discuss specific aspects of hybridization of metaheuristics. Hybrid metaheuristics design, development and testing require a combination of skills and a sound methodology. In particular, comparisons among hybrid techniques and assessment of their performance have to be supported by a sound experimental methodology, and one of the mainstream issues of the workshop is to promote agreed standard experimental methodologies. These motivations are still among the driving forces behind the workshop and, in these four years, we have observed an increasing attention to methodological aspects, from both the empirical and theoretical sides. The papers selected for presentation at HM 2007 are indeed a representative sample of research in the field of hybrid metaheuristics. They range from methodological to application papers. Moreover, some of them put special emphasis on the experimental analysis and statistical assessment of results.

Among the publications in this selection, there are some that focus on the integration of metaheuristics with mathematical programming, constraint satisfaction or machine learning techniques. This interdisciplinary subject is now widely recognized as one of the most effective approaches for tackling hard problems, and there is still room for new results. To achieve them, the community needs to be open toward other research communities dealing with problem solving, such as those belonging to artificial intelligence (AI) and operations research (OR).

We also note that the use of software libraries for implementing metaheuristics is increasing, even though we have to observe that the users of a software library are usually its developers, thus reducing the advantages in terms of software design and development. We believe that this situation is going to change in favor of a scenario in which some libraries will be used by most metaheuristic developers.

Finally, there are also some works describing applications of metaheuristics in continuous optimization. The cross-fertilization between combinatorial and continuous optimization is extremely important, especially because many real-world problems can be naturally modeled as mixtures of discrete and continuous components.

It is already a tradition of the workshop to keep the acceptance rate of papers relatively low: this makes it possible to publish official proceedings, which can be taken as one of the main references in the field. Besides this, special care is taken with respect to the reviewing process, during which the authors are provided with constructive and detailed reviews. For this reason, the role of the Program Committee members is crucial, and we are very grateful to them for the effort they made examining the papers and providing detailed reviews. Among the 37 submissions received, 14 papers have been selected on the basis of the Program Committee members' suggestions. We are further grateful to Catherine C. McGeoch and Thomas Stützle, who both accepted our invitation to give an overview talk.

Looking back to the previous editions of the workshop, we observe a positive trend concerning experimental methodology. Moreover, some topics, such as the integration of metaheuristics with OR and AI techniques, have become established themes. We believe that a grounded discipline in hybrid metaheuristics could bring advantages in problem solving in many areas, such as constrained optimization, mixed integer optimization and also stochastic and online problems, which are probably one of the new frontiers still to be fully explored.

August 2007

Thomas Bartz-Beielstein
María J. Blesa
Christian Blum
Boris Naujoks
Andrea Roli
Günter Rudolph
Michael Sampels

# Organization

## General Chair

Günter Rudolph — Universität Dortmund, Germany

## Program Chairs

| | |
|---|---|
| Thomas Bartz-Beielstein | Fachhochschule Köln, Germany |
| María J. Blesa | Universitat Politécnica de Catalunya, Barcelona, Spain |
| Christian Blum | Universitat Politécnica de Catalunya, Barcelona, Spain |
| Boris Naujoks | Universität Dortmund, Germany |
| Andrea Roli | Università di Bologna, Italy |
| Michael Sampels | Université Libre de Bruxelles, Belgium |

## Local Organization

| | |
|---|---|
| Nicola Beume | Universität Dortmund, Germany |
| Gundel Jankord | Universität Dortmund, Germany |
| Maria Kandyba | Universität Dortmund, Germany |

## Program Committee

| | |
|---|---|
| Dirk Arnold | Dalhousie University, Canada |
| Mauro Birattari | Université Libre de Bruxelles, Belgium |
| Jürgen Branke | AIFB, Universität Karlsruhe, Germany |
| Marco Chiarandini | Syddansk Universitet, Denmark |
| Francisco Chicano | Universidad de Málaga, Spain |
| Carlos Cotta | Universidad de Málaga, Spain |
| Luca Di Gaspero | Università di Udine, Italy |
| Marco Dorigo | Université Libre de Bruxelles, Belgium |
| Michael Emmerich | Universiteit Leiden, The Netherlands |
| Thomas Jansen | Universität Dortmund, Germany |
| Joshua Knowles | University of Manchester, UK |
| Andrea Lodi | Università di Bologna, Italy |
| Vittorio Maniezzo | Università di Bologna, Italy |
| Daniel Merkle | Universität Leipzig, Germany |
| Bernd Meyer | Monash University, Australia |
| Martin Middendorf | Universität Leipzig, Germany |

José A. Moreno            Universidad de La Laguna, Spain
J. Marcos Moreno         Universidad de La Laguna, Spain
David Pelta              Universidad de Granada, Spain
Steven Prestwich         4C, Cork, Ireland
Mike Preuss             Universität Dortmund, Germany
Christian Prins          Université de Technologie de Troyes, France
Günther Raidl           Technische Universität Wien, Austria
Andreas Reinholz         Universität Dortmund, Germany
Andrea Schaerf          Università di Udine, Italy
Marc Sevaux            Université de Bretagne-Sud, France
Kenneth Sörensen         Universiteit Antwerpen, Belgium
Thomas Stützle          Université Libre de Bruxelles, Belgium
Dirk Sudholt            Universität Dortmund, Germany
El-Ghazali Talbi         École Polytechnique Universitaire de Lille,
                        France

## Sponsoring Institutions

Universitat Politécnica de Catalunya, Spain
Deutsche Forschungsgemeinschaft, Germany

# Table of Contents

# Evolutionary Local Search for the Super-Peer Selection Problem and the $p$-Hub Median Problem

Steffen Wolf and Peter Merz

Distributed Algorithms Group
University of Kaiserslautern, Germany
{pmerz,wolf}@informatik.uni-kl.de

**Abstract.** Scalability constitutes a key property in Peer-to-Peer environments. One way to foster this property is the introduction of super-peers, a concept which has gained widespread acceptance in recent years. However, the problem of finding the set of super-peers that minimizes the total communication cost is NP-hard. We present a new heuristic based on Evolutionary Techniques and Local Search to solve this problem. Using actual Internet distance measurements, we demonstrate the savings in total communication cost attainable by such a super-peer topology. Our heuristic can also be applied to the more general Uncapacitated Single Assignment $p$-Hub Median Problem. The Local Search is then further enhanced by generalized *don't look bits*. We show that our heuristic is competitive with other heuristics even in this general problem, and present new best solutions for the largest instances in the well known Australia Post data set.

## 1 Introduction

During recent years *evolutionary algorithms* enhanced with *local search* have been used to solve many NP-hard optimization problems [1,2,3,4]. These heuristics take their power from the problem specific local search, while keeping all favorable features of the evolutionary approach.

We are especially interested in optimization problems connected with topology construction in Peer-to-Peer (P2P) systems. Well-known properties of these fully decentralized P2P systems include self-organizing and fault-tolerant behavior. In contrast to centralized systems, they usually possess neither a single point of failure nor other bottlenecks that affect the entire network at once. However, the scalability of such networks becomes an issue in the case of excessive growth: Communication times tend to increase and the load put on every node grows heavily when the networks get larger. A possible solution to this issue is the introduction of *super-peers*. Super-peers are peers that act as servers for a number of attached common peers, while at the same time, they form a network of equals among themselves. In a super-peer enhanced P2P network, each common peer is attached to exactly one super-peer, which constitutes its link to the remainder of the network. All traffic will be routed via the super-peers [5,6].

T. Bartz-Beielstein et al. (Eds.): HM 2007, LNCS 4771, pp. 1–15, 2007.

To ensure smooth operation, the peers generally wish to maintain low-delay connections to the other peers. Hence, minimum communication cost is the aim when designing super-peer P2P networks. In this paper, we present a heuristic combining local search with evolutionary techniques for the Super-Peer Selection Problem (SPSP), i.e. the problem of finding the set of super-peers and the assignment of all other peers that minimizes the total communication cost.

Our special interest lies in the construction of these P2P overlay topologies. However, the problem of selecting the super-peers is strongly related to a hub location problem: the *Uncapacitated Single Assignment p-Hub Median Problem* (USApHMP) [7]. The USApHMP is a well known optimization problem and has received much attention in the last two decades. With minor adjustments, our heuristic can also be used for the USApHMP, which allows the comparison with other algorithms on established standard test cases.

This paper is organized as follows. In Section 2, we provide an overview of related work. In Section 3, we propose our Super-Peer Selection Heuristic. In Section 4, we present results from experiments on real world Internet data for the Super-Peer Selection Problem, as well as on standard test cases for the USApHMP, and compare the results with those of other recently published algorithms. The paper concludes with an outline for future research in Section 5.

## 2   Related Work

The Super-Peer Selection Problem, as proposed here, has not yet been studied in the literature. However, algorithms designed for the USApHMP can also be used for SPSP. The USApHMP has achieved much attention since it was presented by O'Kelly in [7], along with a set of test cases called CAB. Later, O'Kelly *et al.* also presented means of computing lower bounds for these problems [8]. Exact solutions have been computed by Ernst and Krishnamoorthy for problems with up to 50 nodes in [9]. In this paper, they also introduced a new test set called AP. Ebery presented two more efficient mixed integer linear programs (MILP) for the special case of only 2 or 3 hubs [10], and thus solved a problem with 200 nodes (2 hubs), and a problem with 140 nodes (3 hubs). Also, the authors of [9] presented a Simulated Annealing heuristic that found good solutions for problems with up to 200 nodes.

Skorin-Kapov *et al.* presented TABUHUB [11], a heuristic method based on tabu search. Results were presented only for the smallest problems of the CAB set ($n \leq 25$). Also, neural network approaches have been proposed for the USApHMP. In [12], the memory consumption and the CPU time for these approaches was reduced. However, the neural network approach was again only applied to the smallest problems in the CAB set ($n \leq 15$). Unfortunately, no computation times are given, making comparisons with other heuristics difficult.

The most promising heuristic for the USApHMP so far has been presented by Pérez *et al.* in [13]. It is a hybrid heuristic combining Path Relinking [14] and Variable Neighborhood Search. The heuristic has proven to be very fast with both the CAB and AP sets, faster than any other heuristic. However, it failed to find the optimum in some of the smaller CAB instances and still left room for improvements in the larger instances of the AP set. The local search neighborhoods used in this heuristic differ from the ones used here. Especially, the most expensive neighborhood is missing in [13]. This explains the speed as well as the loss of quality.

Two Genetic Algorithms have been presented by Kratica *et al.* [15]. These GAs are based on different representations and genetic operations. Both feature a mutation operator that favors the assignment of peers to closer super-peers, as well as a sophisticated recombination. The results of the second GA are the best results so far, as they improved the solutions for the larger AP instances found in [13]. However, the approach does not include a local search, and can still be improved. As far as we know, the heuristic we present in this paper is the first heuristic combining evolutionary techniques with local search.

## 3  Super-Peer Selection

When constructing a communication cost efficient and load balanced P2P topology we strive for a topology in which a subset of the nodes will function as super-peers while the rest of the nodes, henceforth called edge peers, is each assigned to one of the super-peers. Adhering to the established properties of super-peer overlay structures, the super-peers are fully connected among themselves and are able to communicate directly with the edge peers assigned to them and with their fellow super-peers. Essentially, the super-peers are forming the core of the network. The edge peers, however, will need to route any communication via their assigned super-peer. An example of such a super-peer topology is shown in Fig. 1. Using a topology of this kind, the communication between edge peers $p_1$ and $p_{11}$ is routed via the super-peers $c_1$ and $c_4$. A broadcast in such a topology can be efficiently performed by having one super-peer send the broadcast to all other super-peers, which then forward the message to their respective edge peers. To ensure smooth operation and to ease the load on each peer, the number of super-peers should be limited as well as the number of peers connected to a super-peer.

The Super-Peer Selection Problem can be defined as finding the super-peer topology, i. e. the set of super-peers and the assignment of the edge peers to the super-peers, with minimal total communication cost for a given network. In a P2P setting, this cost can be thought of as the total all-pairs end-to-end communication delay.

### 3.1  Background

The SPSP is NP-hard [16]. It may be cast as a special case of the Hub Location Problem, first formulated by O'Kelly [7] as a Quadratic Integer Program. In the Hub Location Problem, a number of nodes, the so-called hubs, assume hierarchical superiority over common nodes, a property equivalent to the super-peer concept. Basically, given a network $G = (V, E)$ with $n = |V|$ nodes, $p$ nodes are to be selected as hubs. Let $x_{ik}$ be a binary variable denoting that node $i$

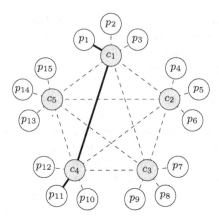

**Fig. 1.** Example of a P2P network with selected Super-Peers

is assigned to node $k$ if and only if $x_{ik} = 1$. If $x_{kk} = 1$, node $k$ is chosen as a hub. The flow volume between any two nodes $i \neq j$ is equal to one unit of flow. Since all flow is routed over the hubs, the actual weight on the inter-hub links is usually larger than one. The transportation cost of one unit of flow on the direct link between nodes $i$ and $j$ amounts to $d_{ij}$. Now, the SPSP formulated as a Hub Location Problem is

$$\min Z = \sum_{i=1}^{n} \sum_{j=1,j\neq i}^{n} \sum_{k=1}^{n} \sum_{m=1}^{n} (d_{ik} + d_{km} + d_{mj}) \cdot x_{ik} \cdot x_{jm} \qquad (1)$$

s. t.

$$x_{ij} \leq x_{jj} \qquad\qquad i,j = 1,\ldots,n \qquad (2)$$

$$\sum_{j=1}^{n} x_{ij} = 1 \qquad\qquad i = 1,\ldots,n \qquad (3)$$

$$\sum_{j=1}^{n} x_{jj} = p \qquad\qquad (4)$$

$$x_{ij} \in \{0,1\} \qquad\qquad i,j = 1,\ldots,n \qquad (5)$$

Equation (1) yields the total communication cost $Z$. The set of constraints (2) ensures that nodes are assigned only to hubs, while (3) enforces the allocation of a node to exactly one hub. Due to constraint (4), there will be exactly $p$ hubs.

A more general formulation uses a demand matrix $W = (w_{ij})$. Here, $w_{ij}$ denotes the flow from node $i$ to $j$ in flow units. Also, special discount factors can be applied for the different edge types. Flow between hubs is subject to a discount factor $0 \leq \alpha \leq 1$, flow from a node to its hub is multiplied by a factor $\delta$, and flow from a hub to a common node is multiplied by a factor $\chi$. The total communication cost $Z$ is then:

$$Z = \sum_{i=1}^{n}\sum_{j=1}^{n}\sum_{k=1}^{n}\sum_{m=1}^{n} (\chi \cdot d_{ik} + \alpha \cdot d_{km} + \delta \cdot d_{mj}) \cdot w_{ij} \cdot x_{ik} \cdot x_{jm} \qquad (6)$$

The distinction between the different types of edges is motivated by an application in the area of mail transport. Here, the distribution cost differs from the collection cost. Also, the transportation cost between the hubs is assumed to be lower since more efficient means of transport can be used for the accumulated amount of flow. This extension might also be applied in the case of communication networks, especially when asymmetric links are considered. However, the most important difference from the SPSP is the introduction of demand factors $w_{ij}$, as will be shown in Section 3.3.

Since the objective function in both programs is quadratic and nonconvex, no efficient way to compute the minimum is known. The usual approach is to transform the problem into a Mixed Integer Linear Program (MILP). A straightforward linearization uses $\mathcal{O}(n^4)$ variables. We resort to an MILP formulation using as few as $\mathcal{O}(n^3)$ variables [17]:

$$\min Z = \sum_{i=1}^{n}\sum_{k=1}^{n} (\chi \cdot O_i + \delta \cdot D_i) \cdot d_{ik} \cdot x_{ik} + \sum_{i=1}^{n}\sum_{k=1}^{n}\sum_{l=1}^{n} \alpha \cdot d_{kl} \cdot y_{ikl} \qquad (7)$$

s. t. (2), (3), (4), (5),

$$\sum_{l=1}^{n} (y_{ikl} - y_{ilk}) = O_i \cdot x_{ik} - \sum_{j=1}^{n} w_{ij} \cdot x_{jk} \qquad i,k = 1,\ldots,n \qquad (8)$$

$$y_{ikl} \geq 0 \qquad\qquad i,k,l = 1,\ldots,n \qquad (9)$$

Here, $O_i = \sum_{j=1}^{n} w_{ij}$ is the outgoing flow for node $i$ and $D_j = \sum_{i=1}^{n} w_{ij}$ is the demand of node $j$. Both values can be calculated directly from the problem instance. The variables $y_{ikl}$ denote the flow volume from hub $k$ to hub $l$ which has originated at peer $i$. Constraints (8) and (9) ensure flow conservation at each node.

An MILP formulation for the SPSP can be derived by fixing $\chi = \delta = \alpha = 1$, $w_{ij} = 1$ for $i \neq j$, $w_{ii} = 0$, and thus $O_i = D_i = n - 1$:

$$\min Z = \sum_{i=1}^{n}\sum_{k=1}^{n} 2 \cdot (n-1) \cdot d_{ik} \cdot x_{ik} + \sum_{i=1}^{n}\sum_{k=1}^{n}\sum_{l=1}^{n} d_{kl} \cdot y_{ikl} \qquad (10)$$

s. t. (2), (3), (4), (5), (9),

$$\sum_{l=1}^{n} (y_{ikl} - y_{ilk}) = (n-1) \cdot x_{ik} - \sum_{j=1,j\neq i}^{n} x_{jk} \qquad i,k = 1,\ldots,n \qquad (11)$$

The factor $2 \cdot (n-1)$ for the edge-peer to super-peer links in (10) is the number of connections using this link. It is based on the assumption that every edge peer needs to communicate with all other $n - 1$ peers, and all other peers need to communicate with this edge peer.

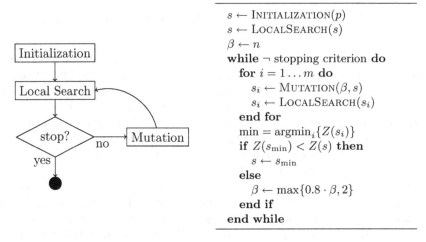

$$s \leftarrow \text{INITIALIZATION}(p)$$
$$s \leftarrow \text{LOCALSEARCH}(s)$$
$$\beta \leftarrow n$$
**while** $\neg$ stopping criterion **do**
    **for** $i = 1 \dots m$ **do**
        $s_i \leftarrow \text{MUTATION}(\beta, s)$
        $s_i \leftarrow \text{LOCALSEARCH}(s_i)$
    **end for**
    $\min = \text{argmin}_i \{Z(s_i)\}$
    **if** $Z(s_{\min}) < Z(s)$ **then**
        $s \leftarrow s_{\min}$
    **else**
        $\beta \leftarrow \max\{0.8 \cdot \beta, 2\}$
    **end if**
**end while**

**Fig. 2.** General overview of the Super-Peer Selection Heuristic

These formulations are equivalent to the quadratic formulation only if the distances $d_{ij}$ observe the triangle inequality. Otherwise, the model will generate solutions featuring the property that messages are sent along shortest paths between two hub instead of the intended direct link. The model can still be used for such networks. However, the resulting value can only serve as a lower bound.

The formulation above enables the exact solution of moderately-sized problems (up to 50 peers) in reasonable time, and additionally, the computation of lower bounds for larger networks (up to 150 peers) using its LP relaxation. For networks larger than the given threshold, we use the lower bounds described in [8]. Finally, the sum of all shortest paths' weights yields another lower bound.

### 3.2   Super-Peer Selection Heuristic

The Super-Peer Selection Heuristic presented here is based on *evolutionary algorithms* and *local search*. It operates on a global view of the network. The general work flow is shown in Fig. 2. The heuristic is quite similar to *iterated local search* [18], but uses more than one offspring solution in each generation.

**Representation.** A solution is represented by the assignment vector $s$. For each peer $i$ the value $s(i)$ represents the super-peer of $i$: $x_{i,s(i)} = 1$. All super-peers, the set of which will be denoted by $C$, are assumed to be assigned to themselves, i.e. $\forall i \in C : s(i) = i$. For the sake of swift computation, we also store the current capacities of the super-peers, i.e. the number of peers connected to the super-peer: $|V_k| = |\{i \in V \mid s(i) = k\}|$. This set also includes the super-peer itself: $k \in V_k$. The sets $V_k$ are not stored explicitly, but are defined by the assignment vector $s$.

**Initialization.** The initial solution is created by randomly selecting $p$ peers as super-peers, and assigning all remaining peers to the nearest super-peer. When handling problems with missing links, this procedure is repeated if the initial set of super-peers is not fully connected.

**Local Search.** After each step of the Evolutionary Algorithm a local search is applied to further improve the current solution. We use three different neighborhoods: *replacing the super-peer, swapping two peers* and *reassigning a peer to another super-peer*. If a neighborhood does not yield an improvement, the next neighborhood is used.

In the first neighborhood the local search tries to replace a super-peer by one of its children. The former child becomes the new super-peer and every other peer that was connected to the old super-peer is reconnected to the new super-peer. The gain of such a move can be computed in $\mathcal{O}(n)$ time. The following formula gives the gain for replacing super-peer $k$ with $i$:

$$G_{\text{replace}}(i, k) = \sum_{j \in C} 2 \cdot |V_k| \cdot |V_j| \cdot (d_{kj} - d_{ij}) + \sum_{j \in V_k} 2 \cdot (n - 1) \cdot (d_{kj} - d_{ij}) \quad (12)$$

If the gain of this move is $G_{\text{replace}}(i, k) > 0$, the move is applied.

The second neighborhood tries to exchange the assignment of two peers. The gain of such a move can be computed in $\mathcal{O}(1)$ time. Since all peers connected to other super-peers are considered as the exchange partner, the total time complexity for searching this neighborhood is $\mathcal{O}(n)$. Using the same notation as before, the following formula gives the gain for swapping the assignments of peers $i$ and $j$:

$$G_{\text{swap}}(i, j) = 2 \cdot (n - 1) \cdot (d_{i,s(i)} + d_{j,s(j)} - d_{i,s(j)} - d_{j,s(i)}) \quad (13)$$

The move with the highest gain is applied if its gain is $G_{\text{swap}}(i, j) > 0$.

The third neighborhood covers the reassignment of a peer to another super-peer. Here, it is important that the capacity limits of the involved super-peers are observed. The gain of reassigning peer $i$ to super-peer $k$ can be calculated in $\mathcal{O}(p)$ time:

$$\begin{aligned} G_{\text{reassign}}(i, k) = \quad & 2 \cdot (n - 1) \cdot (d_{i,s(i)} - d_{i,k}) \\ & + 2 \cdot \left(|V_{s(i)}| \cdot |V_k| - (|V_{s(i)}| - 1) \cdot (|V_k| + 1)\right) \cdot d_{s(i),k} \\ & + 2 \cdot \sum_{j \in C \setminus \{k, s(i)\}} |V_j| \cdot (d_{s(i),j} - d_{k,j}) \end{aligned} \quad (14)$$

The first part of this equation gives the gain on the link between peer $i$ and its super-peer. The second part gives the gain on the link between the old and the new super-peer. The third part gives the gain on all remaining intra-core links. Out of all super-peers only the one with the highest gain is chosen, thus yielding a total time complexity of $\mathcal{O}(p^2)$. The move is applied only if the total gain is $G_{\text{reassign}}(i, k) > 0$.

These local search steps are performed for each peer $i$. The local search is restarted whenever an improving step was found and applied. The local search is thus repeated until no improvement for any peer $i$ can be found, i.e. a local optimum has been reached. Since all peers $i \in V$ are considered in these moves, the time complexity for searching the whole neighborhood is $\mathcal{O}(n^2)$.

**Mutation.** Since local search alone will get stuck in local optima, we use mutation to continue the search. Mutation is done by swapping two random peers. Again, the "gain" of such a swap can be computed by (13). Several mutation steps are applied in each round. The number of mutations is adapted to the success rate. The algorithm starts with $\beta = n$ mutations. If no better solution is found in one generation, the mutation rate $\beta$ is reduced by 20 %. In each round at least two mutations are applied. This way, the algorithm can adapt to the best mutation rate for the individual problem and for the phase of the search. It is our experience that it is favorable to search the whole search space in the beginning, but narrow the search over time, thus gradually shifting exploration to exploitation.

**Population.** Our heuristic uses a population of only one individual. There is no need for recombination. This is mainly motivated by the high computation cost and solution quality of the local search. Using mutation and local search, $m$ offspring solutions are created. The best solution is used as the next generation only if it yielded an improvement. This follows a $(1 + m)$ selection paradigm. If there was no improvement in the $m$ children, the mutation rate $\beta$ is reduced as described before.

**Stopping criterion.** The heuristic is stopped after five consecutive generations without an improvement. This value is a compromise between solution quality and computation time. In the smaller instances the heuristic often finds the optimum in the first or second generation. Continuing the search would mean to waste time. We also stopped the heuristic after 40 generations regardless of recent improvements. Both values were chosen based on preliminary experiments.

### 3.3   Adaptation for the USApHMP

The USApHMP introduces weights $w_{ij}$ on the connections between the nodes. While these weights have been equal for all node pairs in the Super-Peer Selection Problem, this is no longer the case in the full USApHMP. The main effect on the heuristic is that we can no longer summarize the flow on the inter-hub edges as $2 \cdot |V_a| \cdot |V_b|$. The following sum has to be used, instead: $\sum_{i \in V_a} \sum_{j \in V_b} w_{ij} + w_{ji}$. This would change the time complexity for calculating the cost of an inter-hub edge from $\mathcal{O}(1)$ to $\mathcal{O}(n^2)$. With the use of efficient data structures, however, the calculation for the cost of a move can be achieved in $\mathcal{O}(n)$ time.

**Data structures.** In addition to the super-peers' capacities we also store the weights on the $p^2$ inter-hub links. $WC(a,b) = \sum_{i \in V_a} \sum_{j \in V_b} w_{ij}$ denotes the weight on the link from super-peer $a$ to super-peer $b$. In each move made by the local search or the mutation these weights are changed accordingly. Only the selection of a new super-peer does not change these weights. Also, the gain calculations have to be adapted:

$$G_{\text{replace}}(i,k) = \alpha \cdot \sum_{j \in C} (WC(j,k) + WC(k,j)) \cdot (d_{kj} - d_{ij})$$
$$+ \sum_{j \in V_k} (\chi \cdot O_j + \delta \cdot D_j) \cdot (d_{kj} - d_{ij}) \quad (15)$$

$$G_{\text{swap}}(i,j) = \alpha \cdot \sum_{x \in V} (d_{s(j),s(x)} - d_{s(i),s(x)}) \cdot (w_{jx} - w_{ix} + w_{xj} - w_{xi})$$
$$+ (\chi \cdot O_j + \delta \cdot D_j) \cdot (d_{j,s(j)} - d_{j,s(i)})$$
$$+ (\chi \cdot O_i + \delta \cdot D_i) \cdot (d_{i,s(i)} - d_{i,s(j)})$$
$$- 2 \cdot \alpha \cdot d_{s(i),s(j)} \cdot (w_{ij} - w_{ii} + w_{ji} - w_{jj}) \tag{16}$$

$$G_{\text{reassign}}(i,k) = \alpha \cdot \sum_{x \in V} (d_{s(i),s(x)} - d_{k,s(x)}) \cdot (w_{ix} + w_{xi})$$
$$+ (\chi \cdot O_i + \delta \cdot D_i) \cdot (d_{i,s(i)} - d_{ik}) + 2 \cdot \alpha \cdot d_{k,s(i)} \cdot w_{ii} \tag{17}$$

The time complexity of calculating the gains is still $\mathcal{O}(n)$ for replacing the super-peer, but has increased to $\mathcal{O}(n)$ for swapping the assignments of two peers and to $\mathcal{O}(n)$ for reassigning a peer to another super-peer. Using the same local search as presented for the Super-Peer Selection Problem would mean to increase the total time complexity. We therefore implemented a reduced version of the most expensive local search: the swapping of two peers. Instead of calculating the gain for swapping one peer with all other $n - p - 1$ peers, we only calculate this gain for a random sample of $p$ peers, which proved to be a good compromise between computation time and solution quality.

**Don't look markers.** To further speed up the computation we use *don't look markers* to guide the local search. These markers are a generalization of *don't look bits*, that have been applied successfully for example to the *Traveling Salesman Problem* (TSP) [19]. In our algorithm, nodes that do not yield an improvement during one step of the local search will be marked. Nodes that are marked twice or more will not be checked in the following local search steps. However, if a node is part of a successful move all marks are removed again. This can happen if the node is the exchange partner of another node.

Using simple *don't look bits*, i.e. disregarding all nodes with one or more marks, leads to poor results, indeed. Here, too large parts of the considered neighborhoods will be hidden from the local search. Trying to reduce the negative impact of the *don't look bits* immediately lead to the more general *don't look markers*.

## 4  Evaluation

We have performed several experiments to study the effectiveness of our heuristic. For these experiments, we used real world node-to-node delay information from PlanetLab [20], a world-wide platform for performing Internet measurements. We used the round-trip times (RTT) from any host to any other host as the communication cost for the edges in the overlay network. From the measurements reported in [20] we used the first measurement for each month in 2005 (denoted by mm-2005). Those networks consist of $n = 70$ to $n = 419$ nodes. Common properties of all those networks are the frequent triangle inequality violations due to different routing policies and missing links most likely due to firewalls. We strive for $p \approx \sqrt{n}$ super-peers as a viable balance between low load and administrative efficiency. Also, the number $x$ of common peers assigned to a super-peer is limited by $\frac{1}{2}p \leq x \leq 2p$. This ensures that no super-peer suffers from a high load, and that no peer is selected as super-peer without need.

For the USApHMP we used the CAB data set (Civil Aeronautics Board, up to 25 nodes) from [7,21] and the AP data set (Australia Post, up to 200 nodes) from [9]. Both data sets have been widely accepted as standard test cases, and optima for all smaller instances are known. In the AP set the discount factors are fixed to $\alpha = 0.75$, $\chi = 3$ and $\delta = 2$. The CAB set fixes $\chi = \delta = 1$, but provides instances for different $\alpha$-values: 0.2, 0.4, 0.6, 0.8 and 1.0. When refering to the individual instances in these sets, we will use the following notations: AP.$n$.$p$ for the AP instance with $n$ nodes and $p$ hubs, CAB.$n$.$p$.$\alpha$ for the CAB instance with the corresponding configuration, and shorter CAB.$n$.$p$ when $\alpha = 1.0$.

For each problem instance the heuristic was started 30 times and average values are shown. Computation times refer to a 2.8 GHz Pentium 4 with 512 MB RAM running Linux.

### 4.1   Lower Bounds

For networks too large to be handled with the models described in Section 3.1, we are interested in computing lower bounds for SPSP. Table 1 contains the lower bounds for the networks considered here. The *all pairs shortest path* lower bound (APSP) yields the total communication cost (i. e. the sum of the distances of all node pairs) when communication is routed over shortest paths only. Column LB1 holds the lower bound defined in [8].

The column CPlex-LB holds an improved lower bound. We let CPlex 10.1 [22] solve the SPSP and find lower bounds. For the smallest network 04-2005, CPlex required eight days on a 3 GHz Pentium D with 4 GB RAM to arrive at a gap of 1.55 %. For network 01-2005, we stopped CPlex after it has consumed three weeks of CPU time on the same machine. For all other networks, CPlex

**Table 1.** Different lower bounds for the considered networks

| Network | Size | $p$ | APSP | LB1 | CPlex-LB |
|---------|------|-----|------|-----|----------|
| 01-2005 | 127 | 12 | 2 447 260 | 2 501 413 | 2 632 988 |
| 02-2005 | 321 | 19 | 15 868 464 | 16 200 776 | |
| 03-2005 | 324 | 18 | 17 164 734 | 17 580 646 | |
| 04-2005 | 70 | 9 | 663 016 | 690 888 | 728 576 |
| 05-2005 | 374 | 20 | 17 324 082 | 17 794 876 | |
| 06-2005 | 365 | 20 | 18 262 984 | 18 735 944 | |
| 07-2005 | 380 | 20 | 24 867 734 | 25 337 991 | |
| 08-2005 | 402 | 21 | 27 640 136 | 28 151 142 | |
| 09-2005 | 419 | 21 | 23 195 568 | 23 646 759 | |
| 10-2005 | 414 | 21 | 28 348 840 | 28 905 539 | |
| 11-2005 | 407 | 21 | 23 694 130 | 24 196 510 | |
| 12-2005 | 414 | 21 | 20 349 436 | 20 885 442 | |

**Table 2.** Best known solutions, their excess over the LB1 lower bound and the gain compared to random configurations for the considered networks

| Network | Best known | Excess over LB1 | Gain |
|---------|-----------|-----------------|------|
| 01-2005 | 2 929 830 | 17.1 % | 3.8 |
| 02-2005 | 18 620 614 | 14.9 % | 2.7 |
| 03-2005 | 20 715 716 | 17.8 % | 2.7 |
| 04-2005 | 739 954 | 7.1 % | 2.5 |
| 05-2005 | 25 717 036 | 44.5 % | 2.8 |
| 06-2005 | 22 319 228 | 19.1 % | 3.1 |
| 07-2005 | 31 049 398 | 22.5 % | 3.2 |
| 08-2005 | 30 965 218 | 10.0 % | 3.1 |
| 09-2005 | 33 039 868 | 39.7 % | 3.2 |
| 10-2005 | 32 922 594 | 13.9 % | 3.4 |
| 11-2005 | 27 902 552 | 15.3 % | 3.3 |
| 12-2005 | 28 516 682 | 36.5 % | 7.3 |

required an exceedingly long period of time, hence was unable to provide viable results.

### 4.2 Results for the Super-Peer Selection Problem

In Table 2, the best solutions ever found by our heuristic are compared with the lower bounds from the previous table. This also includes runs of the heuristic with higher numbers of offspring and relaxed stopping criterion (500 offsprings, 1000 generations, different $\beta$ adaptation, no stopping after five consecutive generations without improvement). There is still a considerable gap between the LB1 lower bounds and these best known solutions, ranging from 7.1 % to 44.5 %. We believe these best known solutions to be close to the optimum, though, since the LB1 lower bound is too low in instances with many triangle inequality violations. In fact, the best known solutions for both 04-2005 and 01-2005 are still within the lower and upper bounds found by CPlex.

The comparison of these best known solutions with unoptimized super-peer topologies quantifies the benefit of optimization. The column Gain in Table 2 gives the quotient of an average random configuration's cost to the best known solution's cost. Overlay topologies when constructed without locality awareness can be assumed to be random. Accordingly, the communication cost in real world networks becomes subject to reduction by a factor of 2.5 to 3.8 compared to the unoptimized solution, and even the smallest network's total communication cost could still be successfully optimized by a factor of 2.5 compared to its unoptimized counterpart. The high gain of 7.3 in network 12-2005 yields from the large extent of triangle inequality violations in this network.

For the actual parameter settings as described in Section 3.2, Table 3 shows the average excess over the best known solutions, the CPU times per run and the success rate as the number of runs that found the best known solutions. These results show that the heuristic is able to find solutions close to the best known

**Table 3.** Average excess over best known solution and CPU times

| Network | Excess over best known | CPU time per run | # best found |
|---------|------------------------|------------------|--------------|
| 01-2005 | 0.49 % | 13.6 s | 1/30 |
| 02-2005 | 0.72 % | 241.8 s | 1/30 |
| 03-2005 | 0.69 % | 176.9 s | 1/30 |
| 04-2005 | 0.38 % | 1.5 s | 2/30 |
| 05-2005 | 1.06 % | 302.0 s | 0/30 |
| 06-2005 | 0.55 % | 275.7 s | 1/30 |
| 07-2005 | 0.66 % | 299.2 s | 1/30 |
| 08-2005 | 1.00 % | 286.6 s | 0/30 |
| 09-2005 | 1.05 % | 370.8 s | 0/30 |
| 10-2005 | 1.06 % | 384.4 s | 0/30 |
| 11-2005 | 0.83 % | 368.9 s | 0/30 |
| 12-2005 | 0.78 % | 353.8 s | 1/30 |

solutions in all runs. The average excess is never higher than 1.1 %. Even with the tight stopping criterion and the reduced number of offspring the heuristic finds the best known solution in some cases. The CPU times depend on the size of the network. The heuristic could be stopped earlier, but this would result in worse solution quality.

Unfortunately, the results for the SPSP can not be directly compared to other heuristics. However, with the more general USApHMP and established standard test sets we can show that our heuristic is competitive with the algorithms proposed in the literature.

### 4.3  Results for the USApHMP

Since the full USApHMP is more complex than the SPSP, we use the adapted heuristic as described in Section 3.3 for these experiments. The AP set consists of 20 smaller ($n \leq 50$) and 8 larger instances ($n \geq 100$). Optimal solutions are known only for the smaller instances. The CAB set consists of 60 instances (four different sizes up to 25 nodes, three different numbers of hubs, five different discount factors $\alpha$). For all these instances the optimum is known. Again, each experiment was repeated 30 times.

In all 600 runs for the smaller instances of the AP set our heuristic reached the known optimum. Only in six out of all 1800 runs with the CAB set the optimum was not reached. Table 4 gives details on these runs. In all these cases a less strict stopping criterion helps to reach the optimum again.

This very good solution quality can not be kept up on the larger instances of the AP set. Only in 51 out of the 240 runs on these instances the best known solution was reached. However, the average excess above those best known solutions is still considerably good, as Table 5 shows. The best solutions known so far for the larger instances of the AP set have been listed in [15]. Our heuristic is able to reach these solutions in all but one problem (AP.200.20), while taking

**Table 4.** Average excess above the optimum for the CAB set

| Instance | | | Excess over | # optimum |
|---|---|---|---|---|
| $n$ | $p$ | $\alpha$ | optimum | found |
| 15 | 2 | 0.6 | 0.038 % | 29/30 |
| 15 | 3 | 1.0 | 0.044 % | 28/30 |
| 20 | 4 | 1.0 | 0.004 % | 28/30 |
| 25 | 4 | 1.0 | 0.040 % | 29/30 |
| all other | | | — | 30/30 |

**Table 5.** Average excess above the best known solutions for the AP set for $n \geq 100$

| Instance | | Excess over best | # best | Best known | |
|---|---|---|---|---|---|
| $n$ | $p$ | known solution | found | cost | |
| 100 | 5 | 0.00 % | 30/30 | 136 929.444 | |
| 100 | 10 | 0.30 % | 5/30 | 106 469.566 | |
| 100 | 15 | 0.75 % | 1/30 | 90 533.523 | |
| 100 | 20 | 1.62 % | 1/30 | 80 270.962 | |
| 200 | 5 | 0.16 % | 11/30 | 140 062.647 | improved |
| 200 | 10 | 0.17 % | 1/30 | 110 147.657 | |
| 200 | 15 | 0.72 % | 2/30 | 94 459.201 | improved |
| 200 | 20 | 1.35 % | 0/30 | 85 129.343 | |

**Table 6.** Average CPU times per run in seconds for the AP set

| $n$ | $p = 2$ | $p = 3$ | $p = 4$ | $p = 5$ | $p = 10$ | $p = 15$ | $p = 20$ |
|---|---|---|---|---|---|---|---|
| 10 | 0.07 | 0.08 | 0.09 | 0.09 | | | |
| 20 | 0.14 | 0.20 | 0.24 | 0.28 | | | |
| 25 | 0.19 | 0.29 | 0.37 | 0.45 | | | |
| 40 | 0.37 | 0.64 | 0.99 | 1.48 | | | |
| 50 | 0.58 | 0.94 | 1.42 | 2.14 | | | |
| 100 | | | | 11.19 | 25.81 | 63.01 | 77.87 |
| 200 | | | | 58.10 | 188.05 | 305.05 | 417.41 |

roughly the same CPU time as [15]. We have also found new best solutions for two cases (AP.200.5 and AP.200.15). These new best solutions are marked in Table 5. Our heuristic seems to be more effective in finding the best solutions for problems with less hubs. For example the heuristic never failed to find the best solution in AP.100.5 and also often finds the new best solution for AP.200.5. For problems with more hubs the success rate decreases and the average excess over the best known solution increases.

Also, since the search is more complex for problems with more hubs, the average CPU time increases, as Table 6 shows. The computation times range from less than a second for all smaller instances to about seven minutes for the largest one in the AP set. All computation times for the CAB set are well below one second and seem independent on the intra-core discount factor $\alpha$. The larger the problem instance and the more hubs are to be located the more time the heuristic uses. This behavior can also be observed for other heuristics and is therefore not surprising.

## 5   Conclusion

We have presented a hybrid method combining evolutionary algorithms and local search for the Super-Peer Selection Problem and the USApHMP. This heuristic has proven to find optima in all smaller USApHMP instances. The heuristic uses a more thorough search than previous algorithms, and so has found new best solutions for two of the largest instances in the AP set ($n = 200$).

The time complexity improves when applying the heuristic to the Super-Peer Selection Problem. Here, we are able to tackle problems with $n = 400$ and more nodes within reasonable time. The results show that the heuristic is able to optimize the total communication costs in unoptimized real world super-peer topologies by a factor of about 3.

We are also working on a distributed algorithm to solve the SPSP using only local view of the involved peers. First results are already promising. This algorithm will be integrated into a middleware for Peer-to-Peer Desktop Computing.

## References

1. Merz, P., Freisleben, B.: Fitness Landscapes and Memetic Algorithm Design. In: Corne, D., Dorigo, M., Glover, F. (eds.) New Ideas in Optimization, pp. 245–260. McGraw–Hill, London (1999)
2. Merz, P.: Memetic Algorithms for Combinatorial Optimization Problems: Fitness Landscapes and Effective Search Strategies. PhD thesis, Department of Electrical Engineering and Computer Science, University of Siegen, Germany (2000)
3. Hoos, H.H., Stützle, T.: Stochastic Local Search: Foundations and Applications. The Morgan Kaufmann Series in Artificial Intelligence. Morgan Kaufmann, San Francisco (2004)
4. Merz, P., Fischer, T.: A Memetic Algorithm for Large Traveling Salesman Problem Instances. In: MIC 2007. 7th Metaheuristics International Conference (2007)
5. Yang, B., Garcia-Molina, H.: Designing a super-peer network. In: Proceedings of the 19th International Conference on Data Engineering, pp. 49–62 (2003)
6. Li, D., Xiao, N., Lu, X.: Topology and resource discovery in Peer-to-Peer overlay networks. In: Jin, H., Pan, Y., Xiao, N., Sun, J. (eds.) GCC 2004. LNCS, vol. 3252, pp. 221–228. Springer, Heidelberg (2004)
7. O'Kelly, M.E.: A quadratic integer program for the location of interacting hub facilities. European Journal of Operational Research 32, 393–404 (1987)
8. O'Kelly, M.E., Skorin-Kapov, D., Skorin-Kapov, J.: Lower bounds for the hub location problem. Management Science 41(4), 713–721 (1995)
9. Ernst, A.T., Krishnamoorthy, M.: Efficient algorithms for the uncapacitated single allocation $p$-hub median problem. Location Science 4(3), 139–154 (1996)

10. Ebery, J.: Solving large single allocation $p$-hub problems with two or three hubs. European Journal of Operational Research 128(2), 447–458 (2001)
11. Skorin-Kapov, D., Skorin-Kapov, J.: On tabu search for the location of interacting hub facilities. European Journal of Operational Research 73, 502–509 (1994)
12. Domínguez, E., Muñoz, J., Mérida, E.: A recurrent neural network model for the $p$-hub problem. In: Mira, J.M., Álvarez, J.R. (eds.) IWANN 2003. LNCS, vol. 2687, pp. 734–741. Springer, Heidelberg (2003)
13. Pérez, M.P., Rodríguez, F.A., Moreno-Vega, J.M.: A hybrid VNS-path relinking for the p-hub median problem. IMA Journal of Management Mathematics, Oxford University Press (2007)
14. Pérez, M.P., Rodríguez, F.A., Moreno-Vega, J.M.: On the use of path relinking for the $p$-hub median problem. In: Gottlieb, J., Raidl, G.R. (eds.) EvoCOP 2004. LNCS, vol. 3004, pp. 155–164. Springer, Heidelberg (2004)
15. Kratica, J., Stanimirović, Z., Tošić, D., Filipović, V.: Two genetic algorithms for solving the uncapacitated single allocation $p$-hub median problem. European Journal of Operational Research 182(1), 15–28 (2007)
16. Wolf, S.: On the complexity of the uncapacitated single allocation p-hub median problem with equal weights. Internal Report 363/07, University of Kaiserslautern, Kaiserslautern, Germany (2007), available at
    http://dag.informatik.uni-kl.de/papers/Wolf2007SPSP-NP.pdf
17. Ernst, A.T., Krishnamoorthy, M.: An exact solution approach based on shortest-paths for p-hub median problems. INFORMS Journal on Computing 10(2), 149–162 (1998)
18. Lourenço, H.R., Martin, O., Stützle, T.: Iterated Local Search. In: Glover, F., Kochenberger, G. (eds.) Handbook of Metaheuristics, pp. 321–353 (2002)
19. Bentley, J.L.: Experiments on traveling salesman heuristics. In: SODA 1990. Proceedings of the first annual ACM-SIAM symposium on Discrete algorithms, Society for Industrial and Applied Mathematics, Philadelphia, PA, USA, pp. 91–99. ACM Press, New York (1990)
20. Banerjee, S., Griffin, T.G., Pias, M.: The Interdomain Connectivity of PlanetLab Nodes. In: Barakat, C., Pratt, I. (eds.) Proc. of the 5th International Workshop on Passive and Active Network Measurement (2004)
21. O'Kelly, M.E., Bryan, D.L., Skorin-Kapov, D., Skorin-Kapov, J.: Hub network design with single and multiple allocation: A computational study. Location Science 4(3), 125–138 (1996)
22. ILOG S.A.: ILOG CPLEX User's Manual, Gentilly, France, and Mountain View, California (2006), http://www.cplex.com/

# An Effective Memetic Algorithm with Population Management for the Split Delivery Vehicle Routing Problem

Mourad Boudia, Christian Prins, and Mohamed Reghioui*

ICD - OSI, University of Technology of Troyes,
12, Rue Marie Curie, BP 2060, 10010 Troyes France
{mourad.boudia,christian.prins,mohamed.reghioui_hamzaoui}@utt.fr
http://www.utt.fr/labos/LOSI

**Abstract.** This paper studies the Split Delivery Vehicle Routing problem (SDVRP), a variant of the VRP in which multiple visits to customers are allowed. This NP-hard problem is solved by a recent metaheuristic called Memetic Algorithm with Population Management or MA|PM (Sörensen, 2003). It consists in a genetic algorithm, combined with a local search procedure for intensification and a distance measure to control population diversity. Special moves dedicated to split deliveries are introduced in the local search. This solution approach is evaluated and compared with the tabu search algorithm of Archetti et al. (2006) and with lower bounds designed by Belenguer et al. (2000). Our method outperforms the tabu search both in solution quality and running time. On a set of 49 instances, it improves the best-known solution 32 times. The savings obtained confirm the interest and the power of the MA|PM.

**Keywords:** Split delivery vehicle routing problem, Memetic algorithm, Distance measure.

## 1 Introduction and Literature Review

The Vehicle Routing Problem (VRP) is one of the most studied problem in operations research. It consists of servicing customers with known demands, using capacitated vehicles based at a depot-node, to minimize the total cost of the routes. In 2002, Toth and Vigo edited a book entirely devoted to this problem [23]. Today, the best exact method for the VRP is a branch-and-cut-and-price developed by Fukasawa et al. [13]. Efficient metaheuristics for large instances, including memetic algorithms like the one of Prins [18], are surveyed in a book chapter by Cordeau et al. [7].

Contrary to the VRP, each customer may be visited several times in the Split Delivery VRP (SDVRP). The studies on this problem are scarce and relatively

---

* Corresponding author.

T. Bartz-Beielstein et al. (Eds.): HM 2007, LNCS 4771, pp. 16–30, 2007.

recent, even if Dror and Trudeau [9,10] underlined in the 90's that splitting can improve both the routing cost and the number of vehicles. These authors proved some properties of the optimal SDVRP solution and proposed a heuristic. Their tests show that the savings become significant when the average demand per customer exceeds 10% of vehicle capacity. A branch and bound algorithm was developed later by Dror et al. [8], with several integer linear formulations and reduced gaps between lower and upper bounds.

Belenguer et al. [4] presented a better lower bound for the SDVRP, based on a polyhedral approach. They studied a new family of valid inequalities and solved to optimality some instances of literature. We use this new bound to evaluate our memetic algorithm.

In two recent papers, Archetti et al. analyse the conditions in which splitting is interesting [3] and prove that the cost ratio VRP/SDVRP is 2 in the worst case [2]. In [1], they present a tabu search procedure called SplitTabu, which improves the results of Dror and Trudeau [10]. Some extensions of the SDVRP have been investigated. For example, Mullaseril et al. [17] adapt a SDVRP resolution method for a split delivery arc routing problem met in livestock feed distribution. Feillet et al. [11] and Frizzell and Griffin [12] study an SDVRP with simple and multiple time windows.

In this paper, a memetic algorithm with population management or MA|PM [21,22] is developed for the SDVRP. Section 2 states the problem and introduces some notation. The main components and the global structure of the MA|PM are described in section 3, while section 4 is devoted to computational evaluations. A conclusion and some future directions close the paper.

## 2    Problem Statement

The SDVRP is defined on a complete weighted and undirected network $G = (N, E, C)$. $N$ is a set of $n + 1$ nodes indexed from 0 onwards. Node 0 corresponds to a depot with identical vehicles of capacity $W$. Each other node $i$, $i = 1, 2, \ldots, n$, has a known demand $q_i$. The weight $c_{ij} = c_{ji}$ on each edge $(i, j)$ of $E$ is the travelling cost between nodes $i$ and $j$. We assume that no demand $q_i$ exceeds vehicle capacity $W$. Otherwise, for each customer $i$ such that $q_i > W$, an amount of demand $W$ can be deducted from $q_i$ to build one dedicated trip with a full load, until the residual demand fits vehicle capacity, as shown in [1].

Partial deliveries are allowed, so some customers (called *split customers*) can be visited more than once. The objective is to determine a set of vehicle trips of minimum total cost. Each trip starts and ends at the depot and supplies a subset of customers. The number of trips or vehicles used is a decision variable. It is assumed that the triangle inequality holds: in that case, solutions in which each trip visits its customers only once dominate the others. In other words, if one customer is visited several times, these visits are done by distinct trips.

# 3  MA|PM Components

## 3.1  General Principles of MA|PM

In combinatorial optimization, classical genetic algorithms (GA) are not agressive enough, compared to other metaheuristics like tabu search. The memetic algorithms (MA) proposed by Moscato [16] are more powerful versions, in which intensification is performed by applying an improvement procedure (local search) to new solutions. For some classical optimization problems, memetic algorithms are currently the best solution methods. For instance, Prins designed for the VRP one MA which outperforms most published metaheuristics [18].

The MA with population management or MA|PM was introduced by Sörensen in his Ph.D. thesis [21] and recently published in a journal [22]. Like in any incremental MA, starting from an initial population, each iteration selects two parents, applies a crossover operator, improves the offspring using a local search procedure and replaces some existing solutions by the offspring. However, the mutation operator of traditional GAs and MAs is replaced in MA|PM by a diversity control based on a distance measure in solution space.

More precisely, let $d(B, C)$ be the distance between two solutions $B$ and $C$ and, by extension, $D_P(C) = \min\{d(B, C) : B \in P\}$ the distance between a new solution $C$ and a population $P$. $C$ is accepted to replace one solution in $P$ if and only if $D_P(C) \geq \Delta$, where $\Delta$ is a given diversity threshold which can be adjusted during the search. In other words, new solutions enter the population only if they differ enough from existing solutions, in terms of structure.

Sörensen recommends a systematic local search after crossovers. When $D_P(C) < \Delta$, he suggests to apply a mutation operator until $C$ is accepted, but the resulting solution can be strongly degraded. Prins et al. prefer to discard the child in that case, but then the local search rate must be limited to 20-50% to avoid loosing too much time in unproductive iterations. They obtain better results with this option on the Capacitated Arc Routing Problem (CARP) [19].

## 3.2  Chromosome and Evaluation

Like in the VRP MA of Prins [18], each solution is encoded as a list $T$ of customers, without trip delimiters. The list can be viewed as a TSP tour and a procedure called *Split* is required to split it into trips to get a feasible SDVRP solution and its cost. However, in the SDVRP, each customer may appear more than once in $T$ and a parallel list $D$ is required to define the amounts delivered for each visit. Figure 1 shows two examples for an instance with $n = 6$ customers. No demand is split in the first chromosome, while customers 1, 3 and 6 are visited twice in the other. Note that the sum of the amounts delivered to each customer must be equal to his demand, e.g., $1 + 4 = 5$ for customer 3.

The procedure *Split* is easy to understand for the VRP, so we recall it first. An auxiliary graph $H = (X, A, Z)$ is constructed to describe all possible ways of splitting the TSP tour $T$ into trips compatible with vehicle capacity. $X$ contains $n+1$ nodes indexed from 0 onwards. $A$ contains one arc $(i-1, j)$ for each feasible trip (sublist of customers) $(T_i, T_{i+1}, \cdots, T_j)$. The weight $z_{i-1,j}$ of arc $(i-1, j)$

Chromosome $U$ without split demands

List of customer visits $T$: 5 3 6 2 1 4

Amounts delivered $D$ :    9 5 7 4 8 6

Chromosome $V$ with 3 split demands

List of customer visits $T$: 5 3 3 6 6 2 1 1 4

Amounts delivered $D$ :    9 1 4 6 1 4 5 3 6

**Fig. 1.** Two examples of chromosomes with 6 customers

is the cost of the associated trip. The optimal splitting, subject to the sequence imposed by $T$, is obtained by computing a shortest path from node 0 to node $n$ in $H$. Since $H$ contains $O(n^2)$ arcs in the worst case, the shortest path can be computed with the same complexity, using the version of Bellman's algorithm for directed acyclic graphs.

For the SDVRP, due to split demands, $T$ may contain more than $n$ customer visits and a trip $(T_i, T_{i+1}, \ldots, T_j)$ may visit several times some customers, here called *split customers*. Since the triangle inequality holds, the trip cost does not increase if all visits except one are suppressed for each split customer. The problem of setting the best possible cost on the arc which models the trip in $H$ is equivalent to removing superfluous visits to minimize the cost of the resulting trip.

This problem is sometimes easy. For instance, if $T$ contains a trip $(1, 5, 1, 3, 2, 1, 4)$, the only split customer is 1, and there are only three ways of removing redundant visits: keep either the first, second or third occurrence. However, for SDVRP solutions with a high splitting level, the length of $T$ is not bounded by a polynomial in $n$ and the splitting procedure cannot be polynomial. We decided to use a simple rule: for each split customer in a trip, we keep the first visit and delete the other ones. Compared to the VRP the splitting is no longer optimal but can still be implemented in $O(n^2)$. In fact, after some preliminary tests, we realized that using this sub-optimal policy is not a real drawback: in most cases, the local search of the MA is able to move each customer visited by a vehicle to a better position in the trip.

### 3.3   Initial Population

The initial population $P$ contains a fixed number $nc$ of distinct chromosomes. Two are built using VRP heuristics: the Clarke and Wright saving algorithm [6] and the sweep heuristic of Gillett and Miller [15]. The saving algorithm starts from a trivial solution, with one dedicated trip per customer. Then, it performs successive mergers (concatenations of two trips) to reduce the total routing cost, until additional mergers would violate vehicle capacity or increase total cost. In the sweep heuristic, clusters of customers compatible with vehicle capacity are generated by the rotation of a half-line centered on the depot. A vehicle route is then computed in each cluster, by solving a traveling salesman problem.

By construction, no demand is split in these two solutions. Each solution is converted into a chromosome by concatenating its trips. When the average demand per customer is important compared to vehicle capacity, the two resulting chromosomes are often identical and, of course, only one is put in $P$. The $nc - 2$ or $nc - 1$ other initial chromosomes are randomly generated as below, to include a few split demands:

*Step 1.* Generate a random permutation $S$ of the customers.

*Step 2.* Starting from its first customer, $S$ is partitioned into successive trips with a greedy heuristic. Each trip is completed when the vehicle gets full or if adding one more customer would violate vehicle capacity. In the second case, the customer is split to fill the vehicle completely and his remaining demand is used to start a new trip. Hence, only the first and last customers of each trip may be split in the resulting solution. Finally, the trips are concatenated to form the chromosome.

In figure 1, assume that $W = 10$ and that the random permutation generated in step 1 is the list $T$ of chromosome $U$. The heuristic in step 2 extracts one full trip (5,3), one full trip (3,6) in which customer 6 is split, one full trip (6,2,1) in which 6 and 1 are split, and a residual trip (1,4). Once concatenated, these trips yield chromosome $V$.

### 3.4   Selection and Crossover

In each MA iteration, the two parents $A$ and $B$ are chosen using the binary tournament method: two solutions are randomly drawn in $P$ and the best one is taken for $A$. The same process is repeated to get $B$.

The recombination operator is a cyclic, one-point crossover designed for chromosomes with different lengths. Only one child-solution $C$ is generated. The cut-point $k$ is randomly drawn in the integer interval $[1, \min\{|A|, |B|\} - 1]$. Child $C$ is initialized with the first $k$ customers of $A$ and their delivered amounts. The child solution is completed by browsing circularly the second parent $B$, from

Chromosome $A$

Customers : 5 3 4 6|5 2 1 6 4 5
Demands :   3 5 1 3|5 4 8 4 5 1

Chromosome $B$

Customers : 4 1 6 2|5 3 4 1 2 5 6
Demands :   3 3 5 1|6 5 3 5 3 3 2

Chromosome $C$

Customers : 5 3 4 6 5 4 1 2 6 4 1 6 2
Demands :   3 5 1 3 6 3 5 3 2 2 3 2 1

**Fig. 2.** Example of crossover

position $k + 1$ onwards. There are two cases when inspecting one customer $i$. If his demand is already satisfied in $C$, we skip him. Otherwise, the customer is appended to $C$ and the quantity associated to this visit is eventually truncated to guarantee that the total amount he receives in $C$ does not exceed his demand $q_i$. Figure 2 depicts one example of crossover for 6 customers and $k = 4$.

### 3.5   Local Search

After each crossover, child $C$ is evaluated and converted into an SDVRP solution, using the *Split* procedure of 3.2. In this solution, recall that no customer is visited several times by the same trip. The solution is then improved by the local search and reconverted into a chromosome by concatenating its trips. Finally, the distance test described in 3.6 is applied to know if the new chromosome may be accepted. As explained in 3.1, too many children are rejected if the local search is systematic. In practice, it is called with a fixed probability $\beta < 1$.

The first-improvement local search evaluates two groups of moves. The first group does not split demands. The solution to be improved can be viewed as a string with trips separated by copies of the depot node. In the first group, all pairs $(i, j)$ of nodes in this string are considered, *even if they belong to distinct trips*, and the following moves are evaluated if vehicle capacity is respected. The notation $s(i)$ denotes the successor of customer $i$ in his trip.

- move $i$ or $(i, s(i))$ after $j$, if $i$ and $s(i)$ are customers.
- exchange $i$ and $j$ if they are customers.
- 2-opt moves in one trip: the subsequence from $s(i)$ to $j$ is inverted.
- 2-opt moves on two trips: edges $(i, s(i))$ and $(j, s(j))$ are removed and replaced either by $(i, j)$ and $(s(i), s(j))$ or by $(i, s(j))$ and $(j, s(i))$.

These moves are classical for the VRP, but there is a modification for the SD-VRP: if a customer $i$ is visited by a trip and if a second visit to $i$ is moved to this trip, the amount transferred is aggregated with the first visit. This ensures that all stops in a trip concern different customers.

The second group with three moves may split a customer or change the amounts delivered to each visit. The first move is an improvement of the *k-split procedure* of Dror and Trudeau [10] and used later by Archetti et al. [1]. This procedure consists of removing one customer $i$ from all his trips and to reinsert it (possibly with splitting) into a given number of trips $k$, in order to decrease the total routing cost. Note that the best insertion position in a trip is unique and can be pre-computed. Following discussions with these authors, it appears that they only evaluate insertions for small values of $k$, to avoid excessive running times.

We found an optimal method for which $k$ does not need to be fixed. Let $S$ be the set of trips in which $i$ can be inserted (the vehicle is not full), $a_j$ the residual capacity of trip $j$, $u_j$ the fixed cost if $i$ is inserted into $j$, $y_j$ the amount of demand inserted into $j$ and $x_j$ a binary variable equal to 1 if trip $j$ is used for insertion. The optimal reinsertion corresponds to the following integer linear program.

$$\min \sum_{j \in S} u_j x_j \tag{1}$$

$$\sum_{j \in S} y_j = q_i \tag{2}$$

$$0 \leq y_j \leq a_j x_j \qquad \forall j \in S \tag{3}$$

$$x_j \in \{0,1\} \qquad \forall j \in S \tag{4}$$

In fact, this ILP can be replaced by the 0-1 knapsack below. Indeed, constraints (2) and (3) imply (6), the optimal solution $x^*$ of the knapsack problem tells us which trips must be used, and the demand $q_i$ may be assigned as we wish to these trips, e.g., using the very simple algorithm 1.

$$\min \sum_{j \in S} u_j x_j \tag{5}$$

$$\sum_{j \in S} a_j x_j \geq q_i \tag{6}$$

$$x_j \in \{0,1\} \qquad \forall j \in S \tag{7}$$

The knapsack problem can be quickly solved in practice using dynamic programming (DP) and the number of variables (trips) is often much smaller than the number $n$ of customers. We actually implemented the DP method and the MA|PM running time increased by 30%, which is tolerable. However, we decided to use the greedy heuristic in which insertions are done in increasing order of $u_j/a_j$. This heuristic is known to solve optimally the dual of the continuous relaxation of (5)-(7). It is much faster and optimal in 66% of cases for our instances, and we observed no significant increase in the MA|PM final solution cost, compared to the exact algorithm.

---
**Algorithm 1.** Insertion of $i$ using the knapsack problem solution
---
1: **repeat**
2:    $j := \arg\min\{u_p | p \in S \wedge x_p^* = 1 \wedge a_p \neq 0\}$
3:    $z := \min\{q_j, a_j\}$
4:    Insert customer $i$ in trip $j$, with an amount $z$
5:    $q_i := q_i - z; a_j := a_j + z$
6: **until** $q_i = 0$
---

The two last moves are new. The second one depicted in the upper part of figure 3 exchanges two customers $i$ and $j$ pertaining to distinct trips and modify their received amounts. Let $R(i)$, $p(i)$, $s(i)$, and $y(i)$ respectively denote the trip containing one given visit to $i$, the predecessor and successor of $i$ in this trip and the amount received. If $y(i) > y(j)$, $j$ is removed with his amount from $R(j)$ and inserted before or after $i$ in $R(i)$ (the best position is selected). In parallel, a copy of $i$ with an amount $y(i) - y(j)$ is created in $R(j)$, to replace the visit to $j$ which has been moved. If $y(i) < y(j)$, the move is analogous, except that the roles of $i$ and $j$ are exchanged. The case $y(i) = y(j)$ is ignored, since this is

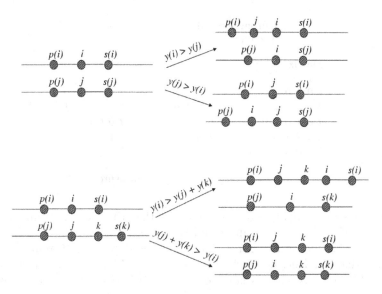

**Fig. 3.** Examples of splitting moves

a standard exchange already treated in the first group. Note that the new move does not affect the load of the trips: in particular, it works on two full vehicles.

The third move is similar but $i$ is swapped with a pair $(j, k)$ of adjacent customers, see the lower part of figure 3. If $y(i) > y(j) + y(k)$, $(j, k)$ is removed from $R(j)$ with its two amounts and inserted before or after $i$ in $R(i)$ (whichever is the best), while a copy of $i$ with a quantity $y(i) - y(j) - y(k)$ is inserted at the old location. If $y(i) < y(j) + y(k)$, by convention, $i$ and $j$ are exchanged without splitting but a copy of $k$ is moved to $R(i)$, with an amount $y(j) + y(k) - y(i)$. Finally, whenever $y(i) = y(j) + y(k)$, a simple exchange of $i$ with $(j, k)$ is performed.

## 3.6  Distance Measure

Campos et al. [5] proposed what they call the "distance for R-permutations". For two permutation chromosomes $A$ and $B$, it is equal to the number of pairs of adjacent customers in $A$ which are broken (no longer adjacent) in $B$. An adaptation of this distance is proposed here to handle SDVRP chromosomes $A$ and $B$ with different lengths. In this case, the distance is the number of pairs present in chromosomes $A$ and not in $B$ plus the number of pairs present in $B$ and not in $A$. Two $n \times n$ matrices $M^A$ and $M^B$ are used. For any chromosome $X$, $M_{ij}^X$ is equal to the number of pairs $(i, j)$ of adjacent customers in $X$. Since we have symmetric edge costs, a trip is equivalent to its reverse trip, i.e. performed backward by the vehicle. To make our distance reversal-independent, $M^X$ is symmetric, i.e. we set $M_{ij}^X = M_{ji}^X = 1$ if $X$ contains the substring $(i, j)$.

The distance is then defined as $d(A, B) = \sum_{i=1}^n \sum_{j=1}^n |M_{ij}^A - M_{ij}^B|/2$. Figure 4 gives one example. The sum of the elements in matrix $|A - B|$, divided by

Chromosome $A$     Chromosome $B$

1 2 4 3 2 4 5      1 4 3 2 5 4

| Matrix $M^A$ | Matrix $M^B$ | $|M^A - M^B|$ |
|---|---|---|
| 1 2 3 4 5 | 1 2 3 4 5 | 1 2 3 4 5 |
| 1 0 1 0 0 0 | 1 0 0 0 1 0 | 1 0 1 0 1 0 |
| 2 1 0 1 2 0 | 2 0 0 1 0 1 | 2 1 0 0 2 1 |
| 3 0 1 0 1 0 | 3 0 1 0 1 0 | 3 0 0 0 0 0 |
| 4 0 2 1 0 1 | 4 1 0 1 0 1 | 4 1 2 0 0 0 |
| 5 0 0 0 1 0 | 5 0 1 0 1 0 | 5 0 1 0 0 0 |

**Fig. 4.** Example of distance measure

2, gives a distance equal to 5. The reason is that the pair $(1,2)$ and the two pairs $(2,4)$ of $A$ do not exist in $B$ and the pairs $(1,4)$ and $(2,5)$ of $B$ do not exist in $A$. This metric is a generalization of the distance of Campos et al. It is reversal-independent, i.e. $d(A, B) = 0$ if $A = B$ or if $B$ is the mirror of $A$. Indeed, if no customer is split, our distance is equal to twice the one of Campos.

### 3.7 Algorithm Overview

The general structure of the MA|PM can now be summarized in algorithm 2. The initial population $P$ includes the two good VRP solutions obtained with the Clarke and Wright heuristic and the sweep heuristic. The distance measure is used to avoid adding a solution $S$ already in $P$, by checking that $d_P(S) \neq 0$. The resulting population is sorted in ascending order of costs.

The main loop performs a fixed number of cycles, *maxcycles* and returns the best solution $P_1$ at the end. Each cycle initializes the diversity threshold $\Delta$, executes a given number of basic iterations *maxcross* and modifies the population for the next cycle, using a partial renewal procedure. Each basic iteration chooses two parents $A$ and $B$ using the binary tournament technique. The resulting child $C$ undergoes the local search with a given probability $\beta$. The solution $R$ to be replaced by $C$ is pre-selected at random in the worst half of $P$. $C$ is accepted if its distance to $P$ minus $R$ is no smaller than $\Delta$. Note that $R$ is not included in the test, because it will no longer be in $P$ in case of replacement. To avoid missing a new best solution, $C$ is also accepted if it improves the current best solution, i.e., if $cost(C) < cost(P_1)$. If $C$ is accepted, it replaces $R$ and a simple shift is sufficient to keep $P$ sorted. Otherwise $C$ is simply discarded. The distance threshold $\Delta$ is updated at the end of each basic iteration.

A decreasing policy is used to vary $\Delta$. Starting from an initial value $\Delta_{max}$, $\Delta$ is linearly decrease to reach 1 (the minimum value to avoid duplicate solutions) at the end of each cycle. The renewal procedure consists of keeping the best solution and of replacing the others by new random solutions. Like in the initial population, all solutions must be distinct.

---

**Algorithm 2.** MA|PM overview

---

1: $P \leftarrow \emptyset$
2: initialize $P_1, P_2$ using the Clarke and Wright and Gillet and Miller heuristics
3: **for** $k = 3$ to $nc$ **do**
4:     **repeat**
5:         $S \leftarrow random\_solution()$
6:     **until** $d_P(S) \geq 1$
7:     $P_k \leftarrow S$
8: **end for**
9: sort $P$ in ascending cost order
10: **for** $cycle = 1$ to $maxcycles$ **do**
11:     initialize diversity threshold $\Delta$
12:     **for** $cross = 1$ to $maxcross$ **do**
13:         select two parent-solutions $A$ and $B$ in $P$ using binary tournament
14:         $C \leftarrow crossover(A, B)$
15:         **if** $random < \beta$ **then**
16:             $C \leftarrow local\_search(C)$
17:         **end if**
18:         select $R$ to be replaced in the worst half of $P$ ($P(\lfloor nc/2 \rfloor)$ to $P(nc)$)
19:         **if** $(D_{P \setminus \{R\}}(C) \geq \Delta)$ or $(cost(C) < cost(P_1))$ **then**
20:             remove solution $R$: $P \leftarrow P \setminus \{R\}$
21:             add solution $C$: $P \leftarrow P \cup \{C\}$
22:             shift $C$ to keep population $P$ sorted
23:         **end if**
24:         update $\Delta$
25:     **end for**
26:     $P \leftarrow partial\_renewal(P)$
27: **end for**
28: return the best solution ($P_1$)

---

# 4   Computational Results

All algorithms designed in this paper were implemented in Delphi and executed on a 3.0 GHz PC with Windows XP. Some preliminary testing was required to tune the parameters, the objective being a good tradeoff between solution quality and running time. The best results on average were achieved using a population size $nc = 10$, four cycles of 2000 crossovers each ($maxcycles = 4$ and $maxcross = 2000$, and a local search rate $\beta = 0.1$. The diversity threshold $\Delta$ is initialized with a $\Delta_{max}$ equal to one quarter of the average inter-solution distance calculated on the initial population. $\Delta$ is then decreased by $(\Delta_{max} - 1)/maxcross$ after each crossover. The tuning seems robust, since it gives very good results on the three sets of instances tested.

The MA|PM was first compared with the tabu search *SplitTabu* of Archetti et al. [1], using the same instances. These instances are derived from the VRP files 1 to 5, 11 and 12 used by Gendreau et al. [14]. They contain 50 to 199 customers. The demands $q_i$ were regenerated randomly for each instance, using six different intervals with respect to vehicle capacity $W$: $[0.01W, 0.1W]$, $[0.1W,$

**Table 1.** Computational results of the SDVRP instances of Archetti

| File | Demand | SplitTabu | | MA|PM | | Saving | Best |
|------|--------|-----------|------|-------|------|--------|------|
| | | Cost | Time | Cost | Time | | |
| p1-50 | | 5 335 535 | 13 | **5 246 111** | 8.53 | 1.68 | 5 246 111 |
| p2-75 | | 8 495 410 | 36 | **8 238 883** | 35.72 | 3.02 | 8 238 883 |
| p3-100 | | 8 356 191 | 58 | **8 294 397** | 34.59 | 0.74 | 8 273 926 |
| p4-150 | | 10 698 369 | 389 | **10 423 740** | 103.69 | 2.57 | 10 254 927 |
| p5-199 | | 13 428 515 | 386 | **13 115 937** | 353.84 | 2.33 | 12 929 951 |
| p6-120 | | 10 560 148 | 38 | **10 412 000** | 50.92 | 1.40 | 10 378 753 |
| p7-100 | | 8 253 184 | 49 | **8 195 575** | 42.89 | 0.70 | 8 195 575 |
| | | | | | | | |
| p1-50 | 0.01-0.1 | 4 637 571 | 5 | **4 607 896** | 12.38 | 0.64 | 4 607 896 |
| p2-75 | 0.01-0.1 | 6 052 376 | 13 | **6 000 642** | 18.75 | 0.85 | 5 962 499 |
| p3-100 | 0.01-0.1 | 7 522 012 | 31 | **7 268 076** | 37.12 | 3.38 | 7 268 076 |
| p4-150 | 0.01-0.1 | 8 909 533 | 73 | **8 756 127** | 100.27 | 1.72 | 8 663 116 |
| p5-199 | 0.01-0.1 | 10 562 679 | 526 | **10 187 055** | 356.22 | 3.56 | 10 183 801 |
| p6-120 | 0.01-0.1 | 10 846 959 | 42 | **9 765 688** | 72.98 | 9.97 | 9 765 688 |
| p7-100 | 0.01-0.1 | **6 487 359** | 58 | 6 497 338 | 34.97 | -0.15 | 6 345 694 |
| | | | | | | | |
| p1-50 | 0.1-0.3 | 7 614 021 | 22 | **7 514 139** | 10.22 | 1.31 | 7 410 562 |
| p2-75 | 0.1-0.3 | 10 953 225 | 45 | **10 744 585** | 34.14 | 1.90 | 10 678 012 |
| p3-100 | 0.1-0.3 | 14 248 114 | 96 | **13 928 469** | 78.06 | 2.24 | 13 772 801 |
| p4-150 | 0.1-0.3 | 19 182 459 | 393 | **18 787 090** | 147.89 | 2.06 | 18 750 888 |
| p5-199 | 0.1-0.3 | 23 841 545 | 755 | **23 401 362** | 347.14 | 1.85 | 23 293 715 |
| p6-120 | 0.1-0.3 | 29 187 092 | 143 | **27 203 752** | 144.19 | 6.80 | 27 203 534 |
| p7-100 | 0.1-0.3 | 14 620 077 | 146 | **14 172 757** | 43.27 | 3.06 | 14 157 818 |
| | | | | | | | |
| p1-50 | 0.1-0.5 | 10 086 663 | 28 | **9 883 062** | 12.49 | 2.02 | 9 883 062 |
| p2-75 | 0.1-0.5 | 14 436 243 | 123 | **14 137 965** | 37.38 | 2.07 | 13 985 257 |
| p3-100 | 0.1-0.5 | 18 947 210 | 136 | **18 452 995** | 28.39 | 2.61 | 18 276 498 |
| p4-150 | 0.1-0.5 | 26 327 126 | 739 | **25 616 472** | 224.89 | 2.70 | 25 397 546 |
| p5-199 | 0.1-0.5 | 32 844 723 | 2 668 | **31 912 475** | 436.20 | 2.84 | 31 802 981 |
| p6-120 | 0.1-0.5 | 42 061 210 | 268 | **39 343 880** | 163.14 | 6.46 | 39 343 880 |
| p7-100 | 0.1-0.5 | 20 299 948 | 293 | **19 945 947** | 51.31 | 1.74 | 19 815 453* |
| | | | | | | | |
| p1-50 | 0.1-0.9 | 14 699 221 | 61 | **14 670 635** | 21.42 | 0.19 | 14 438 367* |
| p2-75 | 0.1-0.9 | 21 244 269 | 193 | **21 025 762** | 46.11 | 1.03 | 20 872 242 |
| p3-100 | 0.1-0.9 | 27 940 774 | 649 | **27 809 492** | 84.38 | 0.47 | 27 467 515* |
| p4-150 | 0.1-0.9 | **39 097 249** | 2 278 | 40 458 715 | 244.91 | -3.48 | 38 497 320* |
| p5-199 | 0.1-0.9 | **48 538 254** | 3 297 | 49 412 170 | 725.69 | -1.80 | 47 374 671* |
| p6-120 | 0.1-0.9 | 65 839 735 | 878 | **63 183 734** | 196.14 | 4.03 | 62 596 720* |
| p7-100 | 0.1-0.9 | **31 015 273** | 260 | 31 137 187 | 52.13 | -0.39 | 30 105 041* |
| | | | | | | | |
| p1-50 | 0.3-0.7 | 14 969 009 | 49 | **14 770 135** | 24.53 | 1.33 | 14 770 135 |
| p2-75 | 0.3-0.7 | 21 605 050 | 129 | **21 321 601** | 51.78 | 1.31 | 21 321 601 |
| p3-100 | 0.3-0.7 | 28 704 954 | 810 | **28 588 684** | 100.16 | 0.41 | 27 642 538* |
| p4-150 | 0.3-0.7 | **40 396 994** | 3 008 | 40 458 715 | 244.86 | -0.15 | 39 671 062* |
| p5-199 | 0.3-0.7 | **51 028 379** | 3 566 | 51 553 604 | 749.94 | -1.03 | 50 014 512* |
| p6-120 | 0.3-0.7 | 66 395 522 | 659 | **64 247 101** | 271.39 | 3.24 | 63 994 197 |
| p7-100 | 0.3-0.7 | **30 380 225** | 778 | 31 556 915 | 91.31 | -3.87 | 28 821 235* |
| | | | | | | | |
| p1-50 | 0.7-0.9 | 21 652 085 | 106 | **21 543 510** | 22.91 | 0.50 | 21 483 778* |
| p2-75 | 0.7-0.9 | **31 806 415** | 869 | 32 003 533 | 27.48 | -0.62 | 31 381 780* |
| p3-100 | 0.7-0.9 | **43 023 114** | 1 398 | 43 129 461 | 55.75 | -0.25 | 42 788 332* |
| p4-150 | 0.7-0.9 | **61 963 577** | 10 223 | 62 674 827 | 401.62 | -1.15 | 60 998 678* |
| p5-199 | 0.7-0.9 | **79 446 339** | 21 849 | 80 815 768 | 571.70 | -1.72 | 76 761 141* |
| p6-120 | 0.7-0.9 | 103 040 778 | 1 826 | **100 634 739** | 298.08 | 2.34 | 100 174 729 |
| p7-100 | 0.7-0.9 | **48 677 857** | 1 004 | 49 194 826 | 180.11 | -1.06 | 47 735 921* |

0.3$W$], [0.1$W$, 0.5$W$], [0.1$W$,0.9$W$], [0.3$W$, 0.7$W$] and [0.7$W$, 0.9$W$]. Including the original instances for which the demands are not changed, the 49 instances listed in Table 1 are obtained. The first group of 7 instances contains the original demands.

The four first columns correspond respectively to the file name with the number $n$ of customers, the interval of demands, the average solution cost found by *SplitTabu* over 5 runs (using different random initial solutions) and the average running time per run in seconds. *SplitTabu* was implemented in C++ on a 2.4 GHz PC. The MA|PM solution value and running time (for one run and the standard setting of parameters) are given in columns 5 and 6. The last but one column indicates the saving in % obtained by our method, computed using the following formula:

$$Saving = \frac{SplitTabu\ cost - MA|PM\ cost}{SplitTabu\ cost} \times 100$$

The last column *Best* gives the new best-known solution values, found either by the tabu search (indicated by an asterisk) or by the MA|PM, when different settings of parameters can be used.

Using one single run, our algorithm improves 37 out of the 49 instances, while being faster than *SplitTabu* for 42. All instances with demands not greater than $W/2$ are improved, except one. The saving for these instances varies between 0.64% and 9.97%. Only instance p7-100 with demands in $[0.01W, 0.1W]$ is not improved, but the loss is marginal (-0.15%). Concerning the other instances, in which some demands may exceed $W/2$, 10 out of 21 are improved. Moreover, MA|PM is much faster than *SplitTabu* on these instances and, when the tabu search is the best, the deviation exceeds 3% only twice: for p4-150 and demands in $[0.1W,0.9W]$ and p7-100 with demands in $[0.3W,0.7W]$. Concerning best-known solutions, 32 are obtained using MA|PM versus 17 for *SplitTabu*.

The second evaluation was conducted with 25 instances used by Belenguer at al. [4], downloadable at http://www.uv.es/~belengue/SDVRP/sdvrplib.html. 11 are VRP instances from the TSPLIB [20], the other 14 are randomly generated. The number in the file names denotes the total number of vertices (customers plus depot). For the random instances, the coordinates come from the TSPLIB instances *eil51*, *eil76* and *eil101*. The demands were randomly generated with

**Table 2.** Results for TSPLIB instances

| File | UB | LB | MA\|PM Cost | Time | MA/LB | UB/LB | MA Best |
|------|------|--------|-------|-------|-------|-------|---------|
| *eil22* | 375 | 375.00 | 375 | 4.11 | 0.00 | 0.00 | 375 |
| *eil23* | 569 | 569.00 | 569 | 5.47 | 0.00 | 0.00 | 569 |
| *eil30* | 510 | 508.00 | **503** | 5.70 | -0.98 | 0.39 | 503 |
| *eil33* | 835 | 833.00 | 835 | 5.19 | 0.24 | 0.24 | 835 |
| *eil51* | 521 | 511.57 | 521 | 7.28 | 1.81 | 1.81 | 521 |
| *eilA76* | 832 | 782.70 | **828** | 35.94 | 5.79 | 6.30 | 818 |
| *eilB76* | 1 023 | 937.47 | **1 019** | 13.09 | 8.70 | 9.12 | 1 007 |
| *eilC76* | **735** | 706.01 | 738 | 14.75 | 4.53 | 4.11 | 733 |
| *eilD76* | 683 | 659.43 | **682** | 23.12 | 3.42 | 3.57 | 682 |
| *eilA101* | **817** | 793.48 | 818 | 25.25 | 3.09 | 2.96 | 815 |
| *eilB101* | **1 077** | 1 005.85 | 1 082 | 21.81 | 7.57 | 7.03 | 1 007 |
| Average | | | | | 3.10 | 3.23 | |

different intervals, as in Archetti et al. [1]. In the names of the randomly gener-
ated instances in Table 3, the code Dx corresponds to the interval used: D1 for
$[\lceil 0.01W \rceil, \lfloor 0.1W \rfloor]$, D2 for $[\lceil 0.1W \rceil, \lfloor 0.3W \rfloor]$, D3 for $[\lceil 0.1W \rceil, \lfloor 0.5W \rfloor]$, D4 for
$[\lceil 0.1W \rceil, \lfloor 0.9W \rfloor]$, D5 for $[\lceil 0.3W \rceil, \lfloor 0.7W \rfloor]$ and D6 for $[\lceil 0.7W \rceil, \lfloor 0.9W \rfloor]$.

Table 2 gives the results obtained for the 11 VRP instances from the TSPLIB.
The first column mentions the file name of each instance, the second and third
give an upper bound $(UB)$ and a lower bound $(LB)$ obtained by Belenguer et
al. using a heuristic method and a cutting plane algorithm. Columns 4 and 5
indicate the solution cost and running time in seconds achieved by our algorithm
with the standard setting of parameters. Column 6 provides the deviation of our
method to the lower bound in %, while the gap $UB/LB$ achieved by Belenguer et
al. is given in column 7 for comparison. The last column gives the best MA|PM
solution found when parameters may be modified.

The average deviation to the lower bound is only slightly improved (around
3%), but the MA retrieves the two optima found by Belenguer et al., improves 4
instances and obtains the same results for two others. Belenguer et al. have a better
solution for 3 instances only. Here, the MA|PM is quite fast, with 36 seconds in
the worst case. For the instance *eil30*, the result obtained is better than the LB.
This is due to the method used to compute this lower bound which considers the
minimum number of vehicles. For the instance *eil30*, the MA|PM uses one more
vehicle than the LB. This result does not put the comparison to LB in jeopardy
because the MA|PM finds the same number of vehicles for all remaining instances.

The comparison with the instances randomly generated by Belenguer et al. is
given in Table 3. The same table format is used. For these instances, the average
gap UB/LB is significantly impoved by the MA: 4.97% versus 8.63%. 13 out of
14 instances are improved and the MA is still reasonably fast, with 50 seconds
in the worst case.

**Table 3.** Results for the random SDVRP instances of Belenguer

| File | UB | LB | MA\|PM Cost | Time | MA/LB | UB/LB | MA Best |
|------|-----|------|------|------|------|------|------|
| S51D1 | 458 | 454.00 | **458** | 8.77 | 0.88 | 0.88 | 458 |
| S51D2 | 726 | 676.63 | **707** | 7.44 | 4.48 | 7.30 | 706 |
| S51D3 | 972 | 905.22 | **945** | 7.84 | 4.39 | 7.38 | 945 |
| S51D4 | 1 677 | 1 520.67 | **1 578** | 11.98 | 3.77 | 10.28 | 1 578 |
| S51D5 | 1 440 | 1 272.86 | **1 351** | 16.72 | 6.14 | 13.13 | 1 336 |
| S51D6 | 2 327 | 2 113.03 | **2 182** | 9.92 | 3.26 | 10.12 | 2 177 |
| S76D1 | 594 | 584.87 | **592** | 15.23 | 1.21 | 1.56 | 592 |
| S76D2 | 1 147 | 1 020.32 | **1 089** | 30.50 | 6.73 | 12.41 | 1 087 |
| S76D3 | 1 474 | 1 346.29 | **1 427** | 12.89 | 5.99 | 9.49 | 1 420 |
| S76D4 | 2 257 | 2 011.64 | **2 117** | 8.76 | 5.24 | 12.15 | 2 094 |
| S101D1 | **716** | 700.56 | 717 | 49.75 | 2.35 | 2.20 | 716 |
| S101D2 | 1 393 | 1 270.97 | **1 372** | 31.72 | 7.95 | 9.60 | 1 372 |
| S101D3 | 1 975 | 1 739.66 | **1 891** | 33.98 | 8.70 | 13.53 | 1 891 |
| S101D5 | 2 915 | 2 630.43 | **2854** | 18.66 | 8.50 | 10.82 | 2 854 |
| Average | | | | 4.97 | 8.63 | | |

# 5   Conclusion and Future Directions

In this paper, the very recent MA|PM framework is applied to the very hard Split Delivery VRP. This requires the design of non-trivial components: an ad-hoc encoding, a crossover operator, an effective local search able to split customers and a distance adapted to chromosomes with varying lengths.

A comparison with Archetti et al. indicates that the proposed algorithm competes with a state of the art tabu search method, while being much faster. The lower bounds available for Belenguer's instances show that solution gaps can be reduced, especially for the randomly generated instances.

Future possible research directions include the design of more effective moves when the average demand represents a large fraction of vehicle capacity, in order to improve the instances which resist. We also intend to tackle additional constraints like time windows. The study of stochastic demands seems very interesting, because demand variations could have a strong impact on a planned solution with split customers. Finally, the deviations to lower bounds are moderate but more important than the typical gaps reachable for the VRP. This raises the need for tighter bounds.

**Acknowledgement.** This research has been funded by the Champagne-Ardenne Regional Council and by the European Social Fund.

# References

1. Archetti, C., Hertz, A., Speranza, M.G.: A tabu search algorithm for the split delivery vehicle routing problem. Transportation Science 40(1), 64–73 (2006)
2. Archetti, C., Savelsbergh, M.W.P., Speranza, M.G.: Worst-case analysis for split delivery vehicle routing problems. Transportation Science 40(2), 226–234 (2006)
3. Archetti, C., Savelsbergh, M.W.P., Speranza, M.G.: To split or not to split: that is the question. Transportation Research Economics (to appear)
4. Belenguer, J.M., Martinez, M.C., Mota, E.: A lower bound for the split delivery vehicle routing problem. Operations Research 48(5), 801–810 (2000)
5. Campos, V., Martí, R., Laguna, M.: Context-independent scatter search and tabu search for permutation problems. INFORMS Journal on Computing 17, 111–122 (2005)
6. Clarke, G., Wright, J.W.: Scheduling of vehicles from a central depot to a number of delivery points. Operations Research 12, 568–581 (1964)
7. Cordeau, J.F., Gendreau, M., Hertz, A., Laporte, G., Sormany, J.S.: New heuristics for the vehicle routing problem. In: Langevin, A., Riopel, D. (eds.) Logistic systems: design and optimization, pp. 279–298. Wiley, Chichester (2005)
8. Dror, M., Laporte, G., Trudeau, P.: Vehicle routing with split deliveries. Discrete Applied Mathematics 50, 239–254 (1994)
9. Dror, M., Trudeau, P.: Savings by split delivery routing. Transportation Science 23, 141–145 (1989)
10. Dror, M., Trudeau, P.: Split delivery routing. Naval Research Logistics 37, 383–402 (1990)
11. Feillet, D., Dejax, P., Gendreau, M., Gueguen, C.: Vehicle routing with time windows and split deliveries. In: Proceedings of Odysseus 2003, Palermo (May 2003)

12. Frizzell, P.W., Giffin, J.W.: The split delivery vehicle scheduling problem with time windows and grid network distances. Computers and Operations Research 22(6), 655–667 (1995)
13. Fukasawa, R., Lysgaard, J., de Aragão, M.P., Reis, M., Uchoa, E., Werneck, R.F.: Robust branch-and-cut-and-price for the capacitated vehicle routing problem. Mathematical Programming 106, 491–511 (2006)
14. Gendreau, M., Hertz, A., Laporte, G.: A tabu search heuristic for the vehicle routing problem. Management Science 40, 1276–1290 (1994)
15. Gillett, B.E., Miller, L.R.: A heuristic algorithm for the vehicle dispatch problem. Operations Research 22, 340–349 (1974)
16. Moscato, P.: Memetic algorithms: a short introduction. In: Corne, D., Dorigo, M., Glover, F. (eds.) New ideas in optimization, pp. 219–234. McGraw-Hill, New York (1999)
17. Mullaseril, P.A., Dror, M., Leung, J.: Split-delivery routing heuristics in livestock feed distribution. Journal of Operational Research Society 48, 107–116 (1997)
18. Prins, C.: A simple and effective evolutionary algorithm for the vehicle routing problem. Computers and Operations Research 31, 1985–2002 (2004)
19. Prins, C., Sevaux, M., Sörensen, K.: A genetic algorithm with population management (GA|PM) for the CARP. In: 5th Triennal Symposium on Transportation Analysis (Tristan V), Le Gosier, Guadeloupe (2004)
20. Reinelt, G.: Tsplib-travelling salesman problem library. ORSAJ. Comput. 3, 376–384 (1991)
21. Sörensen, K.: A framework for robust and flexible optimisation using metaheuristics with applications in supply chain design. PhD thesis, University of Antwerp, Belgium (2003)
22. Sörensen, K., Sevaux, M.: MA|PM: memetic algorithms with population management. Computers and Operations Research 33, 1214–1225 (2006)
23. Toth, P., Vigo, D.: The vehicle routing problem. SIAM, Philadelphia (2002)

# Empirical Analysis of Two Different Metaheuristics for Real-World Vehicle Routing Problems

Tonči Carić[1,*], Juraj Fosin[1], Ante Galić[1], Hrvoje Gold[1],
and Andreas Reinholz[2]

[1] Faculty of Transport and Traffic Sciences, University of Zagreb
Vukelićeva 4, HR-10000 Zagreb, Croatia
{Tonci.Caric,Juraj.Fosin,Ante.Galic,Hrvoje.Gold}@fpz.hr
[2] University of Dortmund, D-44221 Dortmund, Germany
Andreas.Reinholz@gmx.de

**Abstract.** We present two hybrid Metaheuristics, a hybrid Iterated Local Search and a hybrid Simulated Annealing, for solving real-world extensions of the Vehicle Routing Problem with Time Windows. Both hybrid Metaheuristics are based on the same neighborhood generating operators and local search procedures. The initial solutions are obtained by the Coefficient Weighted Distance Time Heuristics with automated parameter tuning. The strategies are compared in an empirical study on four real-world problems. A performance measure is used that also considers multiple restarts of the algorithms.

**Keywords:** Vehicle Routing Problems with Time Windows, Coefficient Weighted Distance Time Heuristics, Iterated Local Search, Simulated Annealing.

## 1 Introduction

The Vehicle Routing Problem (VRP) is defined by the task of finding optimal routes used by a group of vehicles when serving a group of customers. The solution of the problem is a set of routes which all begin and end in the depot, and which suffices the constraint that all the customers are served only once. The objective is to minimize the overall transportation cost. Transportation cost can be improved by reducing the total traveled distance and by reducing the number of the needed vehicles. By adding only capacity constraints to the VRP problem, it is transformed into the most common variation, the Capacitated Vehicle Routing Problem (CVRP). By adding time constraints to the CVRP in the sense that each customer must be served within a customer specific time window, the VRP turns into the well known Vehicle Routing Problem with Time Windows (VRPTW).

The VRPTW is an important NP hard combinatorial optimization problem [1]. It has a wide applicability and has been the subject of extensive research

---

* Corresponding author.

T. Bartz-Beielstein et al. (Eds.): HM 2007, LNCS 4771, pp. 31–44, 2007.

efforts. The solving of real-world VRPTW for delivering or collecting goods needs distance and travel time information between customers that is based on a road network. Therefore the Euclidean metric that is widely used in scientific community for VRPTW (e.g. in the Solomon's benchmarks problems [2]) has to be substituted by real data from an Geographic Information System. Considering the bidirectional nature of traffic flows on streets and road networks the access to this data is organized as an asymmetric look-up matrix. Especially the existence of one-way streets in the cities makes the usage of asymmetric matrices very important for the optimization of routes in the urban area. Due to the time constraints, an additional matrix containing forecasted travel times between each pair of customers has to be created. The quality of forecasting can have a high impact on the feasibility of solutions which are executed in the real-world.

For solving VRPTW problems, a large variety of algorithms has been proposed. Older methods developed for the VRPTW are described in the survey [1] and [3]. Most of the new methods tested on Solomon's benchmarks are comprised in [4], [5].

Methods that applied the two-phase approach of solving VRPTW are found to be the most successful [6]. During the first phase the constructive heuristic algorithm is used to generate a feasible initial solution. In the second phase an iterative improvement heuristic can be applied to the initial solution. The mechanism for escaping from local optimum is often implemented in the second phase, too.

This paper describes the method of finding the strategy that needs less time to produce a solution of desired quality. The success of strategies is determined by the time needed to reach the quality threshold with some probability. Algorithm implementations proposed in the paper could be improved step by step by refining and adding more complex and powerful elements and procedures [7].

Reaching and escaping local optimum are important steps in the process of finding the global optimum for the Iterated Local Search (ILS) and Simulated Annealing (SA). Both strategies are developed in the same computational environment in order to have fair conditions for comparison. Both strategies use the same way of reaching the local optimum which is local search procedure with single $\lambda(1, 0)$ operator for searching the neighborhood [8]. Escaping from local optimum is done by perturbation procedure which is implemented as $k$-step move in the solution neighborhood. The initial solution and the number of iterations are the same as well. Iteration is defined as one cycle of algorithm's outer loop. The second step is significantly different for each strategy. In the applied ILS the perturbated solution is brought to local optimum using the local search procedure. If the new local optimum is better than the so far global best, then that solution is the starting point for a new perturbation; otherwise, the escaping step is discarded. On the contrary, Simulated Annealing never sets the global best solution as the starting point for a new iteration. Also, the series of perturbations are allowed if acceptance criteria are activated successively so that there is a possibility of accepting an inferior solution.

The main contribution of this paper is an attempt to implement the known strategies in the form of simple algorithms on real cases and to conduct the statistical analysis for finding the best suitable strategy for each considered problem. Also, the Coefficient Weighted Distance Time Heuristics (CWDTH) is a novel construction algorithm which gives feasible initial solution for all the considered problems. The automated parameter tuning implemented in CWDTH algorithm enables better adaptation of algorithm to problems with different spatial and temporal distribution of customers. The remainder of the paper is organized as follows: In Section 2 the initial solution methodology and improvement strategies are described. Computational experiments and results are given in Section 3. Finally, in Section 4 conclusions are drawn.

## 2    Solution Methodology

### 2.1    Initialization

In order to solve the VRPTW problems, a constructive heuristic method CWDTH based on the assignment of weights to serving times and distances to the serving places [9] has been developed, Fig. 1.

```
procedure CWDTH()
        for each k[0, 1] in steps of 0.01 do
                s := NewSolution()
                v := FirstVehicle()
                while not Solved()
                        c := BestUnservedCustomer(k)
                        Move(v, c)
                        if CapacityExhausted(v) then
                                Move(v, depot)
                                v := NextVehicle()
                        endif
                endwhile
                remember s if best so far
        next
        return s
end
```

**Fig. 1.** Coefficient Weighted Distance Time Heuristics algorithm

Coefficient interval [0, 1] traversed in empirically determined steps of 0.01 is used for the construction of 101 potentially different solutions. In each pass the algorithm starts from an empty solution and selects the first vehicle. Until all customers are served, the routes are constructed by moving the selected vehicle from its current position to the next best not yet served customer.

Procedure $BestUnservedCustomer()$ uses coefficient $k$ to put different weight to distance and time constraint while selecting the next customer to serve. The criteria of customer selection are:

$$f(x) := MIN(k \cdot Distance(v, c) + (1-k) \cdot LatestTime(c)) \ .$$

Selection of customer $c$ which minimizes function $f(x)$ depends on the sum of its geographic distance to vehicle $v$ multiplied by coefficient $k$ and its upper bound of time window yield by function *LatestTime* multiplied by 1-$k$. After the capacity of the selected vehicle is exhausted, it is returned to the depot to complete the route and the next available vehicle is selected for routing. The best of all the generated solutions is returned as initial solution for further optimisation.

Such approach improves the capability of solving VRPTW problems that have different time window configurations. In other words, the algorithm uses the automated parameter tuning for better adaptation to the specific problem.

## 2.2   Local Search

The local search does not solve the VRP problem from the start, but rather requires in-advance prepared feasible solution obtained by some other method, e. g. CWDTH. The local search generates the neighborhood of the given solution and thus successfully reduces the number of potential solutions that will be candidates for the next iteration.

The mechanism of generating local changes, which is the basis for the success of the iterative local search, is performed by single relocation of the customer from one route into another over the set of all route pairs [8]. On such a way the neighborhood $N_i(s)$ is generated where $s$ stands for the seed solution.

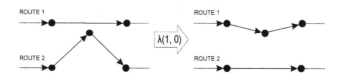

**Fig. 2.** Local search operator $\lambda(1, 0)$

```
procedure LocalSearch(s)
      terminate := false
      do
            find best candidate solution s' in neighborhood N_i(s) produced by λ(1,0)
            if f(s') <f(s) then
                  s := s'
            else
                  terminate := true
            endif
      while not terminate
      return s
end
```

**Fig. 3.** Local search procedure

The principle in which the $\lambda(1,0)$ operator modifies the routes is presented in Fig. 2.

By iterative procedure this local search tries to improve the solution until it is stuck in the local optimum. In each iteration, from the neighborhood of all the feasible moves that respect time and capacity constraints, the best move that produces the most significant saving is chosen to improve the current solution, Fig. 3.

### 2.3   Implementation of Perturbation

Perturbation operator $k$-step move uses the same operator $\lambda(1,0)$ as local search procedure but instead of the best move, random move is chosen to modify the solution, Fig. 4.

```
procedure RandomLocalSearch(s)
      choose random candidate solution s' from neighborhood Nᵢ(s) produced by λ(1,0)
      return s'
end
```

**Fig. 4.** Random local search procedure

This process is $k$ times repeated during one perturbation. The described perturbation gives a feasible solution regarding vehicle capacity and time constraints, Fig. 5.

```
procedure Perturbate(s)
      n := CustomerCount()
      p := 1 / n
      k := BinomialDistribution(p, n)
      for i :=1 to k
            s := RandomLocalSearch(s)
      next
      return s
end
```

**Fig. 5.** Perturbation procedure

The number $k$ is generated by binomial distribution generator with success probability $p$ and number of trials $n$. Values for $p$ and $n$ are empirically obtained. Parameter $n$ is set to number of customers in VRPTW problems and parameter $p$ is set to value $1/n$.

### 2.4   Iterated Local Search

The local search process is started by selecting an initial candidate solution and then proceeded by iteratively moving from one candidate solution to the neighboring candidate solution, where the decision on each search step is based on limited

```
procedure ILS()
    init := CWDTH()
    s := LocalSearch(init)
    best := s
    while not Terminate() do
        s' := Perturbate(s)
        s" := LocalSearch(s')
        if (f(s") < f(best)) then
            best := s"
            s := best
        endif
    endwhile
    return best
end
```

**Fig. 6.** Iterated Local Search Algorithm

amount of local information only. In Stochastic Local Search (SLS) algorithms, these decisions as well as the search initialization can be randomized [10].

Generally, in the Iterated Local Search (ILS) two types of SLS steps are used [11]. One step for reaching local optima as efficiently as possible and the other for efficiently escaping local optima. Fig. 6 shows an algorithm outline for ILS. From the initial candidate solution provided by CWDTH algorithm, local search procedure is performed. Then, each iteration of ILS algorithm consists of three major stages: first, a perturbation is applied to the current candidate solution $s$. This yields a modified candidate solution $s'$ from which in the next stage subsidiary local search procedure is performed until a local optima $s"$ is obtained. In the last stage the new global best solution is updated. The algorithm stops after the termination criterion is met.

## 2.5　Simulated Annealing

Simulated Annealing is a stochastic relaxation technique that finds its origin in statistical mechanics [12], [13], [14]. The name of the method comes from analogy with the annealing process in metallurgy. In the annealing process the material that is heated at high temperature slowly cools and crystallizes under the outside control. Since the heating process allows random movement of atoms, sudden cooling prevents the atoms from achieving the total thermal equilibrium. When the cooling process goes slowly, atoms have enough time to achieve the state of minimal energy forming the ordered crystal grid.

In the optimisation problem the solving of the configuration of atoms is referred to as the state of combinatorial problem, the role of energy is given to cost function and temperature is replaced by control parameter. Simulated Annealing uses stochastic approach to guide the search. The method allows the search to continue in the direction of the neighbor even if the cost function gives inferior results in that direction.

```
procedure SA()
     T := InitialTemperature()
     init := CWDTH()
     s := LocalSearch(init)
     best := s
     while not Terminate() do
          s' := Perturbate(s)
          s" := LocalSearch(s')
          if (f(s") <f(s)) then
               s := s"
          else
               j := rnd(0, 1)
               k := -((f(s")-f(best))/f(best))/T
               if j <exp(k) then
                    s := s"
               endif
          endif
          if (f(s) <f(best)) then
               best := s
          endif
          T := CoolingSchedule()
     endwhile
     return best
end
```

Fig. 7. Simulated Annealing algorithm

In Simulated Annealing algorithm, the starting solution obtained by CWDTH heuristic and local search procedure is the same as for ILS algorithm and it is set as the global best and as the current solution $s$ as well, Fig. 7.

At each iteration of SA the $k$-step perturbation produces solution $s'$. The perturbated solution $s'$ is additionally improved by the local search producing a new solution $s"$. If a new solution $s"$ is better than the current solution $s$, it is accepted as a new current solution. Otherwise, if random generated number within interval $[0, 1)$ is smaller than the current value of acceptance criteria, i.e. $exp(-((f(s")-f(best))/f(best))/T)$, even a worse solution is accepted as the current one. The global best solution is updated if the newly generated solution is better. Initial temperature and cooling schedule are empirically determined once during construction of algorithm and remain the same for all real-world problems and Solomon's benchmark.

## 2.6   Benchmark Results

Before application of algorithms on real-world problems, CWDTH initial solution algorithm and ILS and SA strategies were tested on the standard Solomon's benchmark problems [2]. Comparison of obtained results with the competent results from the literature is shown in Table 1. Testing of both strategies was performed on the 30 independent runs and 4000 iterations as termination criteria.

**Table 1.** Comparison of results obtained by CWDTH, ILS and SA to the best recently proposed results for Solomon's VRPTW problems. CM stands for cumulative values, CPU stands for the processor characteristics and execution time.

| | R1 | R2 | C1 | C2 | RC1 | RC2 | CM | CPU |
|---|---|---|---|---|---|---|---|---|
| HG [15] | 12.08 | 2.82 | 10.00 | 3.00 | 11.50 | 3.25 | 408 | Pentium 400 |
| | 1211.67 | 950.72 | 828.45 | 589.96 | 1395.93 | 1135.09 | 57422 | 3 runs; 1,6 min |
| BC [16] | 12.08 | 2.73 | 10.00 | 3.00 | 11.50 | 3.25 | 407 | Pentium 933 |
| | 1209.19 | 963.62 | 828.38 | 589.86 | 11389.22 | 1143.70 | 57412 | 1 run; 512 min |
| PR [17] | 11.92 | 2.73 | 10.00 | 3.00 | 11.50 | 3.25 | 405 | Pentium 3000 |
| | 1212.39 | 957.72 | 828.38 | 589.86 | 1387.12 | 1123.49 | 57332 | 10 runs; 2,4mi |
| M [18] | 12.00 | 2.73 | 10.00 | 3.00 | 11.50 | 3.25 | 406 | Pentium 800 |
| | 1208.18 | 954.09 | 828.38 | 589.86 | 1387.12 | 1119.70 | 56812 | 1 run; 43,8 min |
| CWDTH | 15.08 | 3.64 | 10.44 | 3.50 | 14.00 | 4.25 | 489 | Centrino 2000 duo |
| | 1543.42 | 1436.86 | 1004.17 | 815.22 | 1797.21 | 1661.52 | 77556 | 100 runs; 0,1 min |
| ILS | 13.67 | 3.55 | 10.00 | 3.25 | 13.25 | 4.25 | 459 | Centrino 2000 duo |
| | 1257.79 | 1022.12 | 839.47 | 613.44 | 1443.52 | 1192.51 | 59888 | 30 runs 7 min |
| SA | 13.08 | 3.27 | 10.11 | 3.25 | 12.63 | 3.75 | 441 | Centrino 2000 duo |
| | 1282.22 | 1053.64 | 901.37 | 621.14 | 1444.15 | 1239.03 | 61523 | 30 runs; 7 min |

It is interesting to observe numerical values of cumulative number of vehicles in Table 1. This number include all variations of Solomon's and roughly depict robustness of algorithms.

Comparison of results obtained by very simple CWDTH constructive algorithm and algorithms with advance techniques for optimization, shows that simplicity degrades results approximatively for 20%. Additional local search with one operator $\lambda(1,0)$ guided by ILS or SA basic strategies can improve solutions for 10% more.

### 2.7   Performance Measure

To compare different algorithms or different parameterized algorithms in an empirical study we were using a performance measure [19] that is motivated by the following question: How often do we have to run an algorithm with a concrete parameter setting so that the resulting solutions are equal or better than a requested quality threshold at a requested accuracy level (i.e. 90%). The lowest number of runs that assures these requests is called multi-start factor ($MSF$). The $MSF$ multiplied by the average runtime of the fixed parameterized algorithm is the performance measure that has to be minimized.

The estimation of the $MSF$ is based on the fact that the success probability $p$ of being better than the requested threshold quality in one run is Bernoulli distributed. Therefore we can use a parameterized maximum likelihood estimator to determine the success probability $p$ for one run. This implies that the success probability for $k$ runs (in $k$ runs there is at least one successful run) is exactly $1 - (1 - p)^k$ and that the $MSF$ for reaching a requested accuracy level($AL$) can be easily computed using a geometrical distribution with success probability $p$

$$MSF := min(k \in \mathbf{N} \quad with \quad 1 - (1 - p)^k \geq AL) \ .$$

When using the intermediate results after each iteration of an algorithm, this procedure can also be used to determine the best combination out of stopping criteria (i.e. maximal number of iterations) and number of restarts for a requested quality threshold and accuracy level.

In this paper we were using the statistical data out of 30 runs for each algorithm and problem instance to estimate the success rates and the average runtimes for all stopping criteria up to 4000 iterations.

The statistical analysis was applied to a series of combinations out of three accuracy levels and two quality thresholds. For the accuracy levels we were using the predefined values 90%, 95%, and 99%. The quality thresholds were chosen out of the data by following procedure: The first quality threshold was defined by the quality value that was reached by the worst out of the 25% best runs after 4000 iterations. The second quality threshold was defined by the quality value that was reached by the worst out of the 10% best runs.

# 3    Computational Results

## 3.1    Characteristics of Data Set

Comparison of ILS and SA strategies was performed on four real-world problems. The objective function for all problems is constructed without penalty because all the operators produce a feasible solution. To force reduction of vehicles in the fleet the objective function is defined as a product of the number of vehicles and the total traveled distance. Problems are classified as VRPTW and described in Table 2.

All problems have homogeneous fleet of vehicles except problem VRP3, and are located in the same geographical area of the city of Zagreb, the capital of Croatia, Fig. 8.

Only one problem, VRP3, spreads on the road networks outside the capital and uses highways between cities. Problems VRP1, VRP3 and VRP4 are mathematically described by distance asymmetric look-up matrix and the related forecasted travel time matrix. The calculation of travel time matrix is based on the average velocity on a particular street or road segments. If such information is not available then the calculation is based on the rank of the road segments. The road ranking follows the national classification which divides them in sixteen categories. Problem VRP2 has linear dependency between the matrix of minimal distances and the matrix of travel times. In problems VRP2 and VRP4

Table 2. Problem characteristics

| Problem | Domain | Customers | Vehicles |
|---------|--------|-----------|----------|
| **VRP1** | Drugs delivery | 64 | 7 |
| **VRP2** | Door-to-door delivery of goods | 90 | 3 |
| **VRP3** | Delivery of consumer goods | 107 | 14 |
| **VRP4** | Newspaper delivery | 154 | 6 |

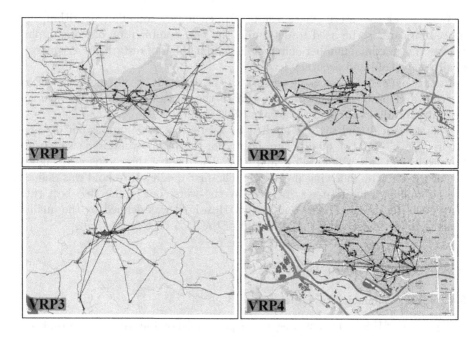

**Fig. 8.** Geographical distribution of customers for all problems

several customers have narrow time windows and in problems VRP1 and VRP3 most of the customers are constrained by terms of working hours of the pickup and delivery department, which means that time windows are relatively wide.

The substitution of the Euclidean metric by the matrix of minimal distances between customers and the use of forecasted travel times for checking time window constraints raises many interesting questions. One of them is the possibility of losing information which can be usable for additional optimisation when we transform real transport networks of streets and roads from geographic information system to asymmetric bidirectional graph with the mentioned asymmetric lookup matrix of minimal distances. At first glance the loss of such information seems to be a problem, but precise analysis leads us to the conclusion that all information that is really important for optimisation of routes are still stored in asymmetric minimal distance matrix. Another important issue is the travel time forecasting model. Such model should be able to predict how much time is needed for a vehicle to move from one geographic location to another in dynamic traffic environment.

## 3.2   Comparative Analysis

Final results of experiments are shown in Table 3 and Table 4. The examination pool of results was constructed by 240 runs of developed ILS and SA algorithms. In order to determine which strategy needs less time, i.e. number of restarts multiplied by the number of iterations, to produce a solution below

**Table 3.** Optimal tuning parameters for VRP problems. $AL$ = Accuracy Level, $T$ = Threshold, $ALG$ = Algorithm, $MSF$ = Multi-start factor, $IT$ = Optimal number of iterations per each run.

| $AL$ | VRP1 | | | VRP2 | | | VRP3 | | | VRP4 | | |
|---|---|---|---|---|---|---|---|---|---|---|---|---|
| 90% | $ALG$ | $MSF$ | $IT$ | $ALG$ | $MSF$ | $IT$ | $ALG$ | $MSF$ | $IT$ | $ALG$ | $MSF$ | $IT$ |
| $T1$ | ILS | 9 | 418 | ILS | 17 | 334 | SA | 9 | 720 | SA | 6 | 1210 |
| $T2$ | ILS | 17 | 1606 | ILS | 17 | 3623 | SA | 34 | 720 | SA | 13 | 1955 |

**Table 4.** Multi-start factors for VRP problems. $AL$ = Accuracy Level, $T$ = Threshold, $IT$ = Optimal number of iterations per each run.

| | VRP1 | | VRP2 | | VRP3 | | VRP4 | |
|---|---|---|---|---|---|---|---|---|
| $AL$ | $T1$ | $T2$ | $T1$ | $T2$ | $T1$ | $T2$ | $T1$ | $T2$ |
| | $IT$=418 | $IT$=1606 | $IT$=334 | $IT$=3623 | $IT$=720 | $IT$=720 | $IT$=1210 | $IT$=1955 |
| 90% | 9 | 17 | 17 | 17 | 9 | 34 | 6 | 13 |
| 95% | 12 | 21 | 21 | 21 | 12 | 44 | 8 | 17 |
| 99% | 26 | 49 | 49 | 49 | 26 | 101 | 18 | 38 |

the threshold with some accuracy, each problem was solved 60 times (each algorithm 30 runs). Table 3 shows optimal parameters of the winning strategy for all problems.

Parameters from Table 3 guarantee reaching of the threshold interval in minimal time with accuracy of 90%. For example, VRP1 needs to be restarted 17 times with halting criteria set to 1606 iterations per start for ILS algorithm. If we increase the accuracy level the number of restarts increases. For each problem the dependencies of multi-start factor and accuracy level are shown in Table 4.

The thresholds are defined in such a way that all the results obtained by ILS and SA are sorted in a list where the value of objective function on the last iteration is the number on which the sorting is done. Threshold $T1$ is calculated so that 25% of runs in the sorted list are in the $T1$ threshold interval. Threshold $T2$ has 10% of best runs. Runs that reached the threshold interval $T2$ for all four problems are depicted in Fig. 9.

Statistical analysis of 30 runs gives us the result for VRP1 and VRP2 revealing that ILS will reach the threshold in less time than SA algorithm. It is reasonable to say that ILS algorithm achieved steeper descent of cost function in fewer iterations for VRP1 and VRP2 compared to SA. On the other hand SA algorithm performs better for two larger problems VRP3 and VRP4. The overall best results for problems VRP3 and VRP4 are obtained by SA near the end of the cooling schedule, so that the resulting graphs confirm the expected convergence behavior of SA at very low temperature. The average running times of algorithms for each problem are given in Table 5. All algorithms are coded in programming language MARS [20].

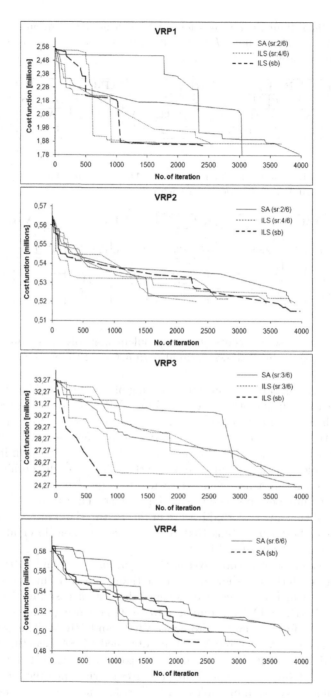

**Fig. 9.** Results that reached threshold interval $T2$, i.e. 10% best runs, for each VRP problem. SA - Simulated Annealing, ILS - Iterated Local Search, sr - success rate, sb - statistically best

**Table 5.** Average running time. The CPU time is given for 10 runs and 4000 iteration per run.

| | VRP1 | | VRP2 | | VRP3 | | VRP4 | |
|---|---|---|---|---|---|---|---|---|
| Algorithm | ILS | SA | ILS | SA | ILS | SA | ILS | SA |
| CPU [min] | 1:03 | 1:02 | 4:55 | 4:46 | 2:01 | 1:54 | 11:31 | 13:13 |

## 4   Conclusions

Test-bed with four real-world VRPTW problems was set up for comparison of two metaheuristic strategies. The Iterated Local Search and the Simulated Annealing strategies are evaluated in the computationally fair environment using the same procedures such as perturbation and local search, and the same experimental setting like the number of iterations and the initial solution.

A decision criterion for choosing the strategy is optimal time for reaching the threshold interval with the targeted accuracy level. The time is represented as the product of the number of runs and the time for reaching maximal iterations per each run. The threshold interval is defined empirically by the number of best runs. The accuracy level is the probability to reach the threshold in the defined time. The results of the conducted experiments give optimal number of restarts and the number of iterations for each run and also refer to the strategy which is best to use to achieve the threshold in minimal time.

The developed algorithms and the applied statistical procedures show that we can validly choose between the well known ILS and SA strategies for a set of test problems. The increase of accuracy level has as a consequence the increase of the number of runs but does not change the number of iterations for this particular set of examined problems.

There is no obvious lead to state that ILS or SA is better than the other one for all the examined problems, but the results show that ILS is better for smaller problem instances and that ILS reaches threshold interval faster. SA works better for larger problems. The application of the described methodology may be able to help a practitioner to roughly approximate the running time with minor changes in topology and constraints of a problem.

The implemented CWDTH algorithm could be a good choice for the initial feasible solution because of its simple implementation especially in the case of practical problems.

New real-world problems with asymmetric matrices of minimal distances and forecasted travel times are introduced.

## References

1. Cordeau, J.-F., Desaulniers, G., Desrosiers, J., Solomon, M., Soumis, F.: The Vehicle Routing Problem with Time Windows. In: Toth, P., Vigo, D. (eds.) The Vehicle Routing Problem, pp. 157–193. SIAM Publishing, Philadelphia (2002)
2. Solomon, M.: Algorithms for the Vehicle Routing and Scheduling Problems with Time Windows Constraints. Operations Research 35, 254–265 (1987)

3. Laporte, G.: The Vehicle Routing Problem: An Overview of Exact and Approximative Algorithms. European Journal of Operational Research 59, 345–358 (1992)
4. Bräysy, O., Gendreau, M.: Vehicle Routing Problem with Time Windows Part I: Route construction and local search algorithms. Trans. Sci. 39, 104–118 (2005)
5. Bräysy, O., Gendreau, M.: Vehicle Routing Problem with Time Windows Part II: Metaheuristics. Trans. Sci. 39, 119–139 (2005)
6. Bräysy, O., Dullaert, W.: A Fast Evolutionary Metaheuristic for the Vehicle Routing Problem with Time Windows. International Journal on Artificial Intelligence Tools 12(2), 153–173 (2003)
7. Reinholz, A.: A Hybrid (1+1)-Evolutionary Algorithm for Periodic and Multiple Depot Vehicle Routing Problems. In: Proceedings on CD. The 6th Metaheuristics International Conference, University of Vienna, Department of Business Administration, Vienna, Austria, August 22-26, 2005, pp. 793–798 (2005)
8. Osman, I.: Metastrategy Simulated Annealing and Tabu Search Algorithms for the Vehicle Routing Problems. Annals of Operation Research 41, 421–451 (1993)
9. Galić, A., Carić, T., Fosin, J., Ćavar, I., Gold, H.: Distributed Solving of the VRPTW with Coefficient Weighted Time Distance and Lambda Local Search Heuristics. In: Biljanović, P., Skala, K. (eds.) Proceedings of the 29th International Convention MIPRO, MIPRO, Rijeka, pp. 247–252 (2006)
10. Hoos, H., Sttzle, T.: Stochastic Local Search: Foundation and Application. Morgan Kauffman, San Francisco (2005)
11. Laurenço, H.R., Serra, D.: Adaptive search heuristics for the generalized assignment problem. Mathware & Soft Computing 9(2-3), 209–234 (2002)
12. Kirkpatrick, S., Gelatt, C.D., Vecchi Jr., M.P.: Optimization by Simulated Annealing. Science 220, 671–680 (1983)
13. Cerny, V.: A Thermodynamical Approach to the Travelling Salesman Problem: An Efficient Simulation Algorithm. Journal of Optimization Theory and Applications 45, 41–51 (1985)
14. Metropolis, N., Rosenbluth, A.W., Rosenbluth, M.N., Teller, A.H.: Equations of State Calculations by Fast Computing Machines. The Journals of Chemical Physics 21, 1087–1092 (1953)
15. Homberger, J., Gehring, H.: A two-phase hybrid metaheuristic for the vehicle routing problem with time windows. European Journal of Operational Research 162, 220–238 (2005)
16. Le Bouthillier, A., Crainic, T.G.: Cooperative parallel method for vehicle routing problems with time windows. Computers and Operations Research 32, 1685–1708 (2005)
17. Pisinger, D., Röpke, S.: A general heuristic for vehicle routing problems. Technical Report, Department of Computer Science, University of Copenhagen (2005)
18. Mester, D., Bräysy, O., Dullaert, W.: A multi-parametric evolution strategies algorithm for vehicle routing problems. Expert Systems with Applications 32, 508–517 (2007)
19. Reinholz, A.: Ein statistischer Test zur Leistungsbewertung von iterativen Variationsverfahren. Technical Report 03027, SFB559, University of Dortmund (2003) (in German)
20. Galić, A., Carić, T., Gold, H.: MARS - A Programming Language for Solving Vehicle Routing Problems. In: Taniguchi, E., Thompson, R. (eds.) Recent Advances in City Logistics, pp. 48–57. Elsevier, Amsterdam (2006)

# Guiding ACO by Problem Relaxation: A Case Study on the Symmetric TSP

Marc Reimann

Institute for Operations Research, ETH Zurich,
Raemistrasse 101, CH-8092 Zurich, Switzerland
Marc.Reimann@ifor.math.ethz.ch

**Abstract.** In this paper the influence of structural information obtained from a problem relaxation on the performance of an ACO algorithm for the symmetric TSP is studied. More precisely, a very simple ACO algorithm is guided by including Minimal Spanning Tree information into the visibility. Empirical results on a large number of benchmark instances from TSPLIB are presented. The paper concludes with remarks on some more elaborate ideas for using problem relaxation within ACO.

## 1 Introduction

The Travelling Salesman Problem (TSP) lies at the heart of many problems in goods distribution, most prominently the Vehicle Routing Problem (VRP). Given a complete graph $G = (V, E)$, where $V$ is a set of $n$ vertices and $E$ is the set of edges, each of which is assigned a weight denoting the edge length, the problem is to find a shortest tour that visits each vertex exactly once and then returns to its origin vertex (for more details see [11]). The TSP is called *symmetric* if the distance between two nodes $i$ and $j$ is the same regardless of the direction of travel, and *asymmetric* otherwise.

Despite this simple problem description, the TSP is an archetypical NP-hard problem, i.e. the number of computations needed to exactly solve an instance of the TSP grows exponentially with the number of vertices to be visited. For more information about the theory of NP completeness see [6].

Consequently, the TSP is often used as a testbed for new algorithmic developments and a large number of heuristic and metaheuristic techniques, e.g. Tabu Search, Simulated Annealing or Ant Colony Optimization (ACO) have first been applied to the TSP.

However, by relaxing some of the TSP constraints one obtains a much simpler problem. Solving this simpler problem yields generally good lower bounds on the objective value of the TSP and the structure of the solution may be used for exact techniques based on, e.g. Branch & Bound. For asymmetric TSPs a very tight bound (usually within 1-2%) is provided by the solution of the assignment problem. Due to that, very large instances of the asymmetric TSP can now be solved to optimality. Also most metaheuristics perform extremely well on these instances.

T. Bartz-Beielstein et al. (Eds.): HM 2007, LNCS 4771, pp. 45–55, 2007.

For symmetric problems, the assignment bound is generally much weaker (due to elementary short cycles) and better bounds can be obtained from the Minimal Spanning Tree (MST), or the 1-Tree solution. Unfortunately, in absolute terms these bounds are not as good as the assignment bounds for the asymmetric TSP and the symmetric TSP is more difficult to solve. On the other hand, when analyzing the structures of the MST and the optimal TSP solution on some benchmark instances it showed that the two solutions have a large number of edges in common (around 70-80%).

Thus, in this paper we will focus on the symmetric TSP and study the influence of exploiting this information within an Ant Colony Optimization (ACO) algorithm. ACO has been first proposed in [2] and is based on stigmergic learning. More precisely, a population of artificial agents repeatedly constructs solutions to the problem using a joint population memory (called pheromone) and some heuristic information (called visibility). After each member of the population has constructed its next solution, the memory is updated with a bias towards better solutions found. Gradually, the memory will build up, thus gaining stronger influence on the solutions built by the artificial agents and the solutions will evolve towards the global optimum. Convergence proofs for generalized versions of ACO algorithms can be found in [8], [9] and [14]. For an overview of recent developments on ACO see [5]. Generally, the best ACO algorithms available today utilize sophisticated local search algorithms and mostly quite elaborate solution construction techniques. While this causes an increase in computational effort per iteration, the obtainable results generally justify this increase.

The approach proposed in this paper deviates from this idea. As we wish to study the influence of using structural insights from the problem relaxation we use a very simple Nearest Neighbor based solution construction for the TSP, where decisions are guided by edge information from the MST. More precisely, edges contained in the MST receive more heuristic weight (referred to as *visibility* in ACO terminology) than those that are not part of the MST. Throughout the paper, this approach will be referred to as *MST-Ants*. Its main advantage is the simplicity. Also the heuristic information is static and can be computed beforehand thus enhancing the speed of the algorithm.

Note that in combining ACO with a lower bounding procedure our approach is related to the algorithm presented in [13], where such a combination was first proposed and applied to the QAP. In that paper each possible move of an ant was evaluated by its impact on the lower bound and moves were discarded when leading to a bound that was larger than the current upper bound. Thus, the bounds were computed dynamically as solutions were built, incorporating the information of the partial solution to be extended. As mentioned above, in our approach the structure of the lower bound solution is exploited only statically by adjusting the visibility once and for all.

The remainder of this paper is organized as follows. In the next section the details of the *MST-Ants* for the TSP are laid out. After that results from a thorough computational study performed on benchmark instances from the TSPLIB are presented before some more ideas for possible utilization of problem relaxation within ACO are discussed.

## 2   *MST-Ants* for the TSP

A high level description of the proposed algorithm is given in Figure 1. Before the iterative part of the ACO algorithm starts an MST is computed using the well known algorithm of Kruskal (c.f. [10]) and the visibility, taking the MST information into account, is then computed beforehand and stored for use within the solution construction of the ACO algorithm.

```
procedure MST-Ants{
    Read the input data;
    Initialize the system (parameters and global pheromone matrix);
    Compute a minimum spanning tree using the algorithm of Kruskal;
    Compute the visibility between all pairs of customers using distance and
    MST information;
    repeat {
        for each ant {
            Construct a TSP tour using a randomized Nearest Neighbor algorithm;
            Apply a 2-opt algorithm to the ants tour;
            Update the best found solution (if applicable);
        }
        Update the global memory (global pheromone matrix);
    } until a pre-specified stopping criterion is met;
    Return the best found solution;
}
```

**Fig. 1.** The *MST-Ants* procedure

The main characteristics of the *MST-Ants*, namely its solution construction, local search and pheromone initialization and update are described in more detail now.

### 2.1   Nearest Neighbor Based Solution Construction

As ACO is a constructive Meta-Heuristic a solution construction mechanism has to be chosen. In most ACO implementations, including the one presented in this paper, this solution construction is based on a Nearest Neighbor mechanism. That is, the ant starts at one of the customers and then sequentially adds customers at the end of its sequence until all the customers have been visited. To avoid multiple visits at customers the ant is equipped with a tabu list, where visited customers are stored. As one generally uses a population of ants much smaller than the problem size, the first question is where to let the ants start. In this paper, each ant is randomly placed at a customer.

The decision making of each ant, i.e. the choice of the next customer to be visited, is based on a probability distribution depending on both the visibility and the pheromone. In a given decision step, let $\Omega$ denote the set of feasible neighbors, i.e. the set of customers that have not yet been visited, and let $i$ be the current location of an ant then the decision rule is given by equation 1, where $\mathcal{P}_{ij}$ is the probability of choosing to move to customer $j$ in the next step.

$$\mathcal{P}_{ij} = \begin{cases} \dfrac{\eta_{ij}\tau_{ij}}{\sum_{h \in \Omega} \eta_{ih}\tau_{ih}} & \text{if } j \in \Omega \\ 0 & \text{otherwise.} \end{cases} \qquad (1)$$

Here $\tau_{ij}$ denotes the pheromone concentration on the edge connecting vertices $i$ and $j$ and contains information from the search history about how good visiting customer $j$ immediately after customer $i$ was in previous iterations.

The visibility $\eta_{ij}$ utilizes heuristic information about the edges. Generally, in the context of the TSP this heuristic information is the inverse distance, i.e.

$$\eta_{ij} = \frac{1}{d_{ij}} \quad \forall (i,j) \in E,$$

where $d_{ij}$ is the distance between customers $i$ and $j$.

The main idea proposed in this paper is to use structural information from the exact solution of a problem relaxation, in this case the MST. Let $t_{ij}$ be a binary variable with $t_{ij} = 1$ if the edge $(i,j)$ is part of the MST solution and $t_{ij} = 0$ otherwise. Then, this information is used to modify the visibility as shown in equation 2, where $\gamma$ is the weight associated to the MST information.

$$\eta_{ij} = \frac{1 + \gamma t_{ij}}{d_{ij}} \quad \forall (i,j) \in E,$$

Clearly, for edges that are not part of the MST the heuristic information is unchanged, while those in the MST will receive higher priority depending on $\gamma$. Note that this approach is static and can be computed once and for all at the beginning of the algorithm. Contrary to that a dynamic approach has been proposed in [13] for the QAP where for each partial solution and each potential move a lower bounding procedure was evoked to compute $\eta_{ij}$. While our approach does not influence runtime crucially, the algorithm in [13] features a significant trade-off between effectiveness and efficiency of the bounding procedure and only simple bounds were used.

## 2.2 Local Search

After the ants have constructed their solutions but before the pheromone is updated each ants solution is improved by applying a local search, more precisely the 2-opt algorithm (c.f. [3]). This algorithm iteratively exchanges two edges with two new edges until no further improvements are possible.

To cut back on the computational effort required for this local search, it is customary to use neighborhood lists for each customer $i$ and to consider only moves for $i$ that include one of its neighbors. However, if the number of neighbors is too small, solution quality deteriorates as some options may not be exploited.

Thus, in the results presented in this paper a neighborhoodsize of $k = n = 4$ is used, where $n$ is the total number of customers in the problem.

### 2.3  Pheromone Initialization and Update

In the constructive phase of the ACO algorithm decisions are based on both heuristic information and the pheromone values as described above. At the end of each iteration, i.e. once all ants have gone through solution construction and local search, these pheromone values are updated as follows:

$$\tau_{ij} := \rho\tau_{ij} + (1 - \rho)\Delta\tau_{ij}^* \quad \forall (i,j) \in E \tag{2}$$

where $0 \le \rho \le 1$ is called the trail persistence and $\Delta\tau_{ij}^*$ is the amount of reinforcement, which is defined as

$$\Delta\tau_{ij}^* := \begin{cases} 1 \text{ if } (i,j) \in S^* \\ 0 \text{ otherwise.} \end{cases} \tag{3}$$

Here $S^*$ is the best solution found up to the current iteration (regardless if it was found in the current iteration or earlier). Note that pheromone values are unit-free and thus independent of monotonous transformations of the objective value. Together with the fact that at the beginning of the run, the pheromone values are initialized to 1, i.e.

$$\tau_{ij} = 1 \quad \forall (i,j) \in E, \tag{4}$$

this update strategy implies that the pheromone values now have a well defined domain, namely $\tau_{ij} \in [0,1]$. In this sense, this approach is similar to the so called Hypercube Framework proposed in [1], where additionally the pheromone values can also be interpreted as transition probabilities which does not hold in our case. However, this interpretation is particularly interesting from a theoretical point of view and less critical when heuristic information is also included as is the case in our approach.

## 3  Numerical Analysis

In this section results of comprehensive computational tests performed to evaluate the performance of the *MST-Ants* are reported. The code was implemented in C and executed on a Pentium III with 900 MHz.

Experiments are based on a set of euclidian problem instances from the TSPLIB.[1] More precisely all instances with less than 1000 customers have been chosen, totalling 47 instances.[2]

To run the *MST-Ants* a number of parameters have to be set. The settings are $m = 20$ Ants, $\rho = 0.975$, $\tau_0 = 1$ and a neighborhoodsize of $k = n/4$. The maximum number of iterations was set to $5n$.

---

[1] The problem data and information about the optimal/best known results can be obtained from the webpage
http://www.iwr.uni-heidelberg.de/iwr/comopt/software/TSPLIB95/

[2] Note that there are actually 48 instances with less than 1000 customers. However, the instance fl417 was not used, due to a problem with the input data.

## 3.1   Effects of Using the MST Information

As the main issue studied in this paper is the question whether the structural information from the problem relaxation is useful for the ants, we will analyze the influence of the MST information first. To that end, we compare results with $\gamma = 1$, i.e. the *MST-Ants*, with the case where $\gamma = 0$ which corresponds to the classic visibility without MST information, referred to as *NN-Ants*. Both approaches were applied 5 times to each problem instance. Note, that it is not the goal of this paper to optimally parameterize the MST information, such that no systematic analysis of different values $\gamma > 0$ was performed.

As the problem instances vary between 51 and 783 customers, the analysis of results will be based on a partition of the instances into two groups depending on the problem size and the computation times required to solve these instances. More precisely, the set of instances comprises 37 instances with at most 318 customers and 10 instances with 400 or more customers. Moreover, for the first group of instances, we will refer to this group as *SMALL*, the average computational effort to reach the best solution varies from 0 to 400 seconds, whereas for the second class of problems, which we will refer to as *LARGE*, it takes between 750 and 5200 seconds to come up with the best solution. Thus, while the partition chosen is somewhat arbitrary, it reflects the differences between the two groups both with respect to size and computational effort quite nicely. Note that complete results for all instances are provided in Appendix 1.

Table 1 gives for both approaches the best, average and worst relative percentage deviation from the optimal solution (RPD) over both groups of instances, as well as the average times (in seconds) to reach the best solution.

The results in Table 1 highlight the differences in the two groups of instances. While the *SMALL* instances can be solved very efficiently by both approaches the *LARGE* instances are much more difficult. First, both the *MST-Ants* and the *NN-Ants* need much larger computation times for reaching good solutions in these instances and second, the solution quality is significantly worse.

Note, that the overall computation time for both approaches is equal as the only difference is the value for $\gamma$, which has no influence on the total complexity of the algorithm. Still, for the *LARGE* instances the actual time needed to find the best solution seems to be better for the *MST-Ants*. Moreover, for these instances the best solution found by the *MST-Ants* seems to be better than the best solution found by the *NN-Ants*. On the other hand, for the *SMALL*

**Table 1.** RPD and computation times needed by the *MST-Ants* and the *NN-Ants* for *SMALL* and *LARGE* instances

| Problem group | MST-Ants RPD (best / avg. / worst) | sec. | NN-Ants RPD(best / avg. / worst) | sec. |
|---|---|---|---|---|
| SMALL | 0.04 / 0.14 / 0.29 | 66.97 | 0.02 / 0.13 / 0.30 | 69.22 |
| LARGE | 0.58 / 0.84 / 1.11 | 2112.03 | 0.67 / 1.00 / 1.35 | 2375.64 |

**Table 2.** Results of the Wilcoxon signed ranks tests for differences in the RPDs and computation times of the *MST-Ants* and the *NN-Ants*

| Problem | best RPD | | average RPD | | worst RPD | | computation time | |
|---------|----|----|-----|----|-----|----|-----|----|
| group | $T$ | $T0$ | $T$ | $T0$ | $T$ | $T0$ | $T$ | $T0$ |
| *SMALL* | 4 | 2 | 144 | 90 | 122 | 73 | 290 | 198 |
| *LARGE* | 11 | 6 | 6* | 8 | 0* | 8 | 19 | 5 |

instances the solution quality and computation times are basically identical for the two approaches.

To validate these results statistically, a Wilcoxon signed ranks test was applied to the samples. More precisely, tests for differences in the best, average and worst RPDs obtained as well as the computation times needed by the two variants, *MST-Ants* and *NN-Ants*, were performed for both the *SMALL* and *LARGE* instances. For each of these cases the test statistic $T$ together with the critical values $T0$ are reported in Table 2. All the results are based on 1-sided tests whether or not the *MST-Ants* outperform the *NN-Ants* and are valid at a 2.5% level of confidence. The null hypothesis can be rejected if $T \leq T0$, i.e. in these cases the *MST-Ants* significantly outperform the *NN-Ants*.

Results marked with an asterisk indicate that the null hypothesis could be rejected. Thus, Table 2 reveals that for the *SMALL* instances there is no significant difference in the RPDs between the *MST-Ants* and the *NN-Ants*. On the other hand, for the *LARGE* instances the inclusion of MST information yields significantly better average and worst case behavior, while it leaves the quality of the best solutions unchanged.

Looking at computation times, Table 1 suggests that the efficiency of the algorithm for *LARGE* instances improves if MST information is used. As can be seen from Table 1, the *MST-Ants* take about 2112 seconds on average to find the best solutions for the *LARGE* instances, while the *NN-Ants* take about 2375 seconds on average to find their best solutions, thus consuming approximately 4.5 minutes more computational effort. However, as can be seen from Table 2, the statistical analysis does not confirm a significant difference in the computation times needed by the two variants for either of the two problem groups.

Summarizing these results, the *MST-Ants* seem to outperform the *NN-Ants* with respect to both quality and runtime on the *LARGE* instances, however only the difference in quality is statistically significant. For the *SMALL* instances there is no difference in a statistical sense. Thus, the use of MST information clearly improves the robustness of the algorithm in terms of the variability of the results as the problem complexity (size) increases.

### 3.2 Comparing *MST-Ants* with Other as Algorithms

From Table 1 it was obvious that the absolute results of the *MST-Ants* differ quite significantly for the two groups of problems analyzed, i.e. the performance of the *MST-Ants* was much worse on the *LARGE* instances than on the *SMALL* instances.

**Table 3.** Comparison of RPDs and computation times needed by the *MST-Ants*, the *GACS* and the *ACS* for *SMALL* and *LARGE* instances

| Problem group | MST-Ants RPD (best/avg.) | sec. | GACS RPD (best/avg.) | sec. | ACS RPD (best/avg.) | sec. |
|---|---|---|---|---|---|---|
| *SMALL* | 0.04 / 0.14 | 150.68 | 0.10 / 0.13 | 493.86 | 0.18 / 0.36 | 296.86 |
| *LARGE* | 0.58 / 0.84 | 4752.07 | 1.36 / 1.46 | 8807.56 | 1.49 / 1.75 | 12459.67 |

The analysis performed in this section will relate the quality obtained by the *MST-Ants* with the quality obtained by two other Ant algorithms presented in [12] to get a more realistic picture about this relationship between size and obtainable solution quality as well as about the performance of the *MST-Ants* in absolute terms.

The two algorithms used for comparison are a standard Ant Colony System (*ACS*) implementation and a combination of the GENI algorithm proposed in [7] with an *ACS*, denoted as *GACS*. Both algorithms were implemented by Le Louarn et al. (c.f. [12]), where the standard *ACS* is a re-implementation of the algorithm proposed in [4].

The main idea of the *GACS* is to replace the Nearest Neighbor based solution construction with the sophisticated GENI algorithm that is based on an insertion criterion. Thus, an unrouted customer may not only be appended at the end of the route, but inserted at its best position along the planned route as this generally improves the solution significantly. More details about this approach can be found in [7] and [12].

Table 3 shows the results of the comparison between the *MST-Ants* and the two other ACO algorithms. As in the last section results are provided over the two classes, *SMALL* and *LARGE*, and the columns show the best and average RPDs and the computation times in seconds for the competing approaches.[3] Note, that complete results for each instance are presented in Appendix 1.[4]

The results for the *GACS* and the standard *ACS* are based on three runs for each instance on an Ultra Sparc 2 with 400 MHz. As stated above, the results for the *MST-Ants* were computed on a Pentium III with 900 MHz. Given these differences in the machines used, the computation times for the *MST-Ants* presented in Table 3 have been multiplied with 2.25 to provide a fair comparison. Note that in Appendix 1 the original computation times are presented.

Table 3 clearly suggests the superiority of the *MST-Ants* over the competing *ACS* algorithms both in terms of solution quality and in terms of computational requirements. Furthermore, both aspects become more striking for the *LARGE* instances where the worst RPD obtained by the *MST-Ants* (given in Table 1) is better than the best RPD by the two competing Ant Systems. Apart from that computation times differ greatly, even after adjusting for machine differences.

---

[3] Worst RPDs are not shown as those are not reported in [12].

[4] Note further, that Table 3 presents averages over 35 *SMALL* and 9 *LARGE* instances, as the instances kroa100, pr299 and u574 have been used for parameter tuning in [12] and have thus been excluded from further testing.

**Table 4.** Results of the Wilcoxon signed ranks tests for differences in the RPDs and computation times of the *MST-Ants*, the *GACS* and the *ACS*

| | MST-Ants vs. GACS | | | | | | MST-Ants vs. ACS | | | | | |
| | best RPD | | avg. RPD | | computation time | | best RPD | | avg. RPD | | computation time | |
| Problem group | $T$ | $T0$ | $T$ | $T0$ | $T$ | $T0$ | $T$ | $T0$ | $T$ | $T0$ | $T$ | $T0$ |
|---|---|---|---|---|---|---|---|---|---|---|---|---|
| SMALL | 7* | 14 | 97 | 73 | 32* | 195 | 17.5* | 59 | 37* | 127 | 161* | 195 |
| LARGE | 0* | 6 | 0* | 6 | 3* | 6 | 0* | 6 | 0* | 6 | 0* | 6 |

Once again, a Wilcoxon signed ranks test was used to statistically support these statements for both the best and the average RPDs and the computation times of the different approaches. Table 4 shows the test statistic $T$ together with the critical values $T0$ for a confidence level of 2.5% for both the *SMALL* and the *LARGE* instances.

As in Table 2, entries marked with an asterisk denote that the null hypothesis of no differences in the samples could be rejected. Thus, Table 4 clearly shows, that the best and the average RPD of the *MST-Ants* significantly outperforms its counterparts of *GACS* and *ACS* for the *LARGE* instances. In fact, for each of the *LARGE* instances both the best and the average RPD are better than the corresponding values of the *GACS* and the *ACS* approach.

For the *SMALL* instances the Wilcoxon signed ranks test shows that the *MST-Ants* outperform the *ACS* for both the best and the average RPD, while the difference between the *MST-Ants* and the *GACS* is statistically significant only for the best RPD, whereas there is no difference for the average RPD.

Looking at computation times, the results in Table 4 show that the *MST-Ants* find their best solutions significantly earlier than both the *GACS* and the *ACS* for both problem groups. Overall, the conclusion is that the *MST-Ants* are clearly superior to both *GACS* and *ACS* with respect to both solution quality and computational effort required.

## 4   Conclusion

In this paper a possible approach for guiding ACO by structural information from the MST relaxation of the symmetric TSP was presented. Within a very simple ACO algorithm the visibility was augmented by the edge information from the MST. It was shown and statistically validated over a large set of benchmark instances that using the MST information improves the robustness of the ACO algorithm significantly. Moreover, particularly for large instances the approach finds solutions of superior quality in smaller computation times when compared with other existing ACO algorithms.

The approach laid out in this paper is one possibility of combining problem relaxation with ACO for the TSP. While the results obtained are very promising, it is worth noting that some other approaches may be even better. Among those are the following issues for further research. First, instead of using the MST information in the visibility it may be better to initialize the pheromone matrix with this information. Particularly, if some information from the MST may be misleading, pheromone initialization could be an interesting opportunity as in this case the tree information is not permanent but rather - through evaporation - can be forgotten over time.

Second, instead of computing the MST once at the beginning of the algorithm it may be favorable to repeatedly compute an MST based on distance data modified by pheromone information. Thus, as the ants build up their pheromone memory, this information could be used to build alternative MSTs and to come up with more diverse and hopefully better solutions.

Third, and this is the most invasive approach, it may be possible to start each ants' solution construction with an MST such that the ant tries to turn the MST into a feasible TSP solution.

Finally, looking at the approach at a more general level reveals that it may be applicable to other combinatorial optimization problems as well. The basic idea is to relax some constraints of the original MILP formulation, to solve the relaxed problem exactly and to use information thus obtained to generate solutions for the original problem with ACO. Even if the problem relaxation is not a known combinatorial problem, as the MST Problem in this case, the fractional decision variables in an LP relaxation may be useful as initial pheromone or visibility for the ants. This approach has also been pointed out in [13].

## Acknowledgments

The author would like to thank three anonymous referees for their valuable comments on the original manuscript which helped to significantly improve the presentation and clarity of the paper.

## References

1. Blum, C., Dorigo, M.: The Hyper-Cube framework for ant colony optimization. IEEE Transactions on Systems, Man and Cybernetics B 34(2), 1161–1772 (2004)
2. Colorni, A., Dorigo, M., Maniezzo, V.: Distributed Optimization by Ant Colonies. In: Varela, F., Bourgine, P. (eds.) Proc. Europ. Conf. Artificial Life, Elsevier, Amsterdam (1991)
3. Croes, G.A.: A method for solving Traveling Salesman Problems. Operations Research 6, 791–801 (1958)
4. Dorigo, M., Gambardella, L.M.: Ant Colony System: A cooperative learning approach to the Travelling Salesman Problem. IEEE Transactions on Evolutionary Computation 1(1), 53–66 (1997)
5. Dorigo, M., Stuetzle, T.: Ant Colony Optimization. MIT Press/Bradford Books, Cambridge, MA (2004)
6. Garey, M.R., Johnson, D.S.: Computers and Intractability: A Guide to the Theory of NP Completeness. W. H. Freeman & Co., New York (1979)
7. Gendreau, M., Hertz, A., Laporte, G.: New insertion and postoptimization procedures for the travelling salesman problem. Operations Research 40, 1086–1094 (1992)

8. Gutjahr, W.J.: A graph-based Ant System and its convergence. Future Generation Computing Systems 16, 873–888 (2000)

9. Gutjahr, W.J.: ACO algorithms with guaranteed convergence to the optimal solution. Information Processing Letters 82, 145–153 (2002)

10. Kruskal, J.B.: On the shortest spanning subtree of a graph and the traveling salesman problem. Proceedings of the American Mathematical Society 7, 48–50 (1956)

11. Lawler, E.L., Lenstra, J.K., Kan, A.H.G.R., Schmoys, D.B. (eds.): The Traveling Salesman Problem. Wiley, Chichester (1985)

12. Le Louarn, F.-X., Gendreau, M., Potvin, J.-Y.: GENI Ants for the travelling salesman problem. Annals of Operations Research 131, 187–201 (2004)

13. Maniezzo, V.: Exact and Approximate Nondeterministic Tree-Search Procedures for the Quadratic Assignment Problem. INFORMS Journal on Computing 11(4), 358–369 (1999)

14. Stuetzle, T., Dorigo, M.: A short convergence proof for a class of ACO algorithms. IEEE Transactions on Evolutionary Computation 6(4), 358–365 (2002)

# Appendix 1

In this appendix, complete results over all problem instances are provided for the *MST-Ants* and the *NN-Ants* proposed in this paper, as well as the *GACS* and the *ACS* from [12]. More precisely, Table 5 contains the following information. The first column gives the names of the problem instances, where the numbers in the names correspond to the problem size, i.e. the number of customers. The second column shows the optimum solutions for these instances. Columns 3 and 4 correspond to the *MST-Ants*. Column 3 provides the best, average and worst RPD (relative percentage deviation over the optimal solution) obtained by the *MST-Ants* over 5 runs for each instance, while column 4 shows the average times required to reach the best results. Columns 5 and 6 show the same performance measures for the *NN-Ants*. Columns 7 and 8 correspond to the *GACS*, where column 7 shows the best and average RPD for the *GACS* obtained over 3 runs for each instance, while column 8 presents the associated computation times. Finally, columns 9 and 10 show the same performance measures for the *ACS*.

**Table 5.** Complete results for all instances with less than 1000 customers, for *MST-Ants*, *NN-Ants*, *GACS* and *ACS*

| Problem instance | Opt. | MST-Ants best/avg./worst | sec. | NN-Ants best/avg./worst | sec. | GACS best/avg. | sec. | ACS best/avg. | sec. |
|---|---|---|---|---|---|---|---|---|---|
| eil51 | 426 | 0.00/0.00/0.00 | 2.64 | 0.00/0.05/0.23 | 1.6 | 0.00/0.00 | 8 | 0.00/0.08 | 1 |
| berlin52 | 7542 | 0.00/0.00/0.00 | 0.38 | 0.00/0.00/0.00 | 0.6 | 0.00/0.00 | 1 | 0.00/0.00 | 1 |
| st70 | 675 | 0.00/0.00/0.00 | 3.78 | 0.00/0.00/0.00 | 3.4 | 0.00/0.00 | 3 | 0.00/0.40 | 26 |
| eil76 | 538 | 0.00/0.00/0.00 | 6.04 | 0.00/0.07/0.19 | 5 | 0.00/0.00 | 80 | 0.00/0.00 | 3 |
| pr76 | 108159 | 0.00/0.00/0.00 | 3.4 | 0.00/0.00/0.00 | 3.8 | 0.00/0.00 | 10 | 0.00/0.00 | 3 |
| rat99 | 1211 | 0.00/0.02/0.08 | 11.33 | 0.00/0.03/0.08 | 7.8 | 0.00/0.00 | 80 | 0.00/0.25 | 137 |
| kroa100 | 21282 | 0.00/0.00/0.00 | 10.58 | 0.00/0.00/0.00 | 9.2 | | | | |
| krob100 | 22141 | 0.00/0.00/0.00 | 10.96 | 0.00/0.10/0.26 | 10.2 | 0.00/0.00 | 135 | 0.00/0.23 | 55 |
| kroc100 | 20749 | 0.00/0.00/0.00 | 9.44 | 0.00/0.00/0.00 | 8.6 | 0.00/0.00 | 8 | 0.00/0.00 | 41 |
| krod100 | 21294 | 0.00/0.03/0.07 | 12.84 | 0.00/0.04/0.07 | 9.8 | 0.00/0.00 | 31 | 0.00/0.06 | 19 |
| kroe100 | 22068 | 0.00/0.05/0.24 | 12.09 | 0.00/0.11/0.33 | 11.2 | 0.00/0.00 | 119 | 0.15/0.33 | 94 |
| rd100 | 7910 | 0.00/0.02/0.08 | 10.96 | 0.00/0.09/0.43 | 10.4 | 0.00/0.00 | 27 | 0.45/0.77 | 9 |
| eil101 | 629 | 0.00/0.16/0.48 | 13.6 | 0.00/0.00/0.00 | 12.6 | 0.00/0.05 | 222 | 0.08/0.64 | 28 |
| lin105 | 14379 | 0.00/0.00/0.00 | 7.56 | 0.00/0.00/0.00 | 6.2 | 0.00/0.00 | 8 | 0.00/0.00 | 9 |
| pr107 | 44303 | 0.00/0.06/0.19 | 12.84 | 0.00/0.00/0.00 | 11.2 | 0.00/0.00 | 19 | 0.30/0.40 | 15 |
| pr124 | 59030 | 0.00/0.00/0.00 | 11.71 | 0.00/0.00/0.00 | 14 | 0.00/0.00 | 36 | 0.00/0.05 | 9 |
| bier127 | 118282 | 0.00/0.00/0.00 | 27.58 | 0.00/0.00/0.00 | 26.6 | 0.00/0.08 | 321 | 0.10/0.48 | 136 |
| ch130 | 6110 | 0.00/0.34/0.74 | 28.33 | 0.00/0.33/0.74 | 27 | 0.00/0.00 | 218 | 0.21/0.45 | 236 |
| pr136 | 96772 | 0.02/0.16/0.46 | 30.22 | 0.01/0.18/0.51 | 27.2 | 0.22/0.28 | 516 | 0.12/0.51 | 104 |
| pr144 | 58537 | 0.00/0.00/0.00 | 18.13 | 0.00/0.00/0.00 | 18.4 | 0.00/0.00 | 375 | 0.00/0.00 | 813.7 |
| ch150 | 6528 | 0.00/0.23/0.43 | 40.42 | 0.00/0.16/0.32 | 37.2 | 0.12/0.28 | 123 | 0.25/0.39 | 21 |
| kroa150 | 26524 | 0.00/0.00/0.00 | 40.8 | 0.00/0.18/0.72 | 41.6 | 0.00/0.00 | 555 | 0.00/0.01 | 661 |
| krob150 | 26130 | 0.00/0.12/0.26 | 44.2 | 0.00/0.19/0.37 | 44.4 | 0.00/0.01 | 254 | 0.05/0.05 | 93 |
| pr152 | 73682 | 0.00/0.07/0.18 | 29.84 | 0.00/0.04/0.18 | 32.8 | 0.00/0.00 | 301 | 0.00/0.06 | 428 |
| u159 | 42080 | 0.00/0.00/0.00 | 19.27 | 0.00/0.00/0.00 | 16.8 | 0.00/0.00 | 142 | 0.19/0.47 | 223 |
| rat195 | 2323 | 0.26/0.60/1.21 | 69.89 | 0.22/0.71/1.03 | 74.4 | 1.16/1.24 | 666 | 1.10/1.44 | 464 |
| d198 | 15780 | 0.02/0.13/0.22 | 95.96 | 0.01/0.06/0.18 | 111.2 | 0.11/0.14 | 1349 | 0.55/0.61 | 229 |
| kroa200 | 29368 | 0.05/0.35/1.13 | 107.67 | 0.00/0.07/0.14 | 127.6 | 0.12/0.18 | 786 | 0.28/0.50 | 72 |
| krob200 | 29437 | 0.07/0.51/0.81 | 103.13 | 0.04/0.26/0.43 | 115.4 | 0.00/0.04 | 1358 | 0.03/0.35 | 1459 |
| ts225 | 126643 | 0.00/0.06/0.13 | 130.71 | 0.00/0.15/0.26 | 112 | 0.06/0.08 | 587 | 0.10/0.18 | 34 |
| tsp225 | 3916 | 0.08/0.21/0.61 | 168.87 | 0.08/0.17/0.43 | 148.4 | 0.00/0.09 | 1046 | 0.40/0.56 | 673 |
| pr226 | 80369 | 0.00/0.07/0.34 | 88.4 | 0.00/0.07/0.29 | 92 | 0.00/0.01 | 1252 | 0.04/0.31 | 840 |
| gil262 | 2378 | 0.04/0.40/0.97 | 307.51 | 0.08/0.45/0.76 | 269 | 0.34/0.41 | 910 | 0.75/0.91 | 155 |
| pr264 | 49135 | 0.00/0.09/0.26 | 117.11 | 0.00/0.05/0.19 | 168.8 | 0.08/0.14 | 1629 | 0.27/0.64 | 674 |
| a280 | 2579 | 0.00/0.09/0.23 | 260.67 | 0.00/0.15/0.47 | 244.4 | 0.38/0.55 | 2400 | 0.38/0.86 | 1514 |
| pr299 | 48191 | 0.09/0.24/0.35 | 308.27 | 0.07/0.34/0.72 | 304 | | | | |
| lin318 | 42029 | 0.68/1.06/1.28 | 312.42 | 0.10/0.86/1.63 | 396.2 | 0.86/0.94 | 1718 | 0.35/0.64 | 1121 |
| rd400 | 15281 | 0.31/0.90/1.28 | 1335.82 | 0.41/1.21/1.68 | 1084.2 | 0.87/0.98 | 6465 | 1.18/1.34 | 4581 |
| pr439 | 107217 | 0.05/0.09/0.20 | 820.91 | 0.05/0.18/0.25 | 743.4 | 0.79/0.84 | 5583 | 1.24/1.75 | 6710 |
| pcb442 | 50778 | 0.62/0.88/1.21 | 964.09 | 0.83/1.10/1.61 | 1042.6 | 1.52/1.58 | 2398 | 1.60/1.99 | 3219 |
| d493 | 35002 | 0.39/0.54/0.69 | 1503.18 | 0.80/1.05/1.27 | 1323.6 | 1.32/1.54 | 9807 | 1.92/2.08 | 4924 |
| u574 | 36905 | 0.56/1.21/1.69 | 1792.18 | 0.08/0.97/1.87 | 2056.8 | | | | |
| rat575 | 6773 | 0.78/1.09/1.48 | 1812.96 | 1.21/1.36/1.62 | 1787.4 | 2.05/2.13 | 12624 | 1.49/1.63 | 11955 |
| p654 | 34643 | 0.08/0.14/0.20 | 2089.49 | 0.10/0.18/0.25 | 2509.8 | 0.34/0.43 | 10879 | 0.72/1.10 | 13582 |
| d657 | 48912 | 0.89/1.06/1.32 | 3582.84 | 0.80/1.12/1.50 | 3170.8 | 1.46/1.61 | 10066 | 1.42/1.50 | 14706 |
| u724 | 41910 | 0.95/1.05/1.14 | 3219.8 | 1.09/1.16/1.31 | 4871 | 1.94/2.05 | 14398 | 1.85/2.26 | 24340 |
| rat783 | 8806 | 1.14/1.39/1.91 | 4098.89 | 1.33/1.68/2.15 | 5166.8 | 1.94/2.02 | 7048 | 1.97/2.11 | 28120 |

# Hybrid Local Search Techniques for the Resource-Constrained Project Scheduling Problem

Igor Pesek[1], Andrea Schaerf[2], and Janez Žerovnik[1,3]

[1] IMFM, Jadranska 19, 1000 Ljubljana, Slovenia
[2] DIEGM, University of Udine, via delle Scienze 208, 33100 Udine, Italy
[3] FS, University of Maribor, Smetanova 17, 2000 Maribor, Slovenia

**Abstract.** This paper proposes a local search algorithm that makes use of a complex neighborhood relation based on a hybridization with a constructive heuristics for the classical resource-constrained project scheduling problem (RCPSP).

We perform an experimental analysis to tune the parameters of our algorithm and to compare it with a tabu search based on a combination of neighborhood relations borrowed from the literature. Finally, we show that our algorithm is also competitive with the ones reported in the literature.

## 1 Introduction

The Resource-Constrained Project Scheduling Problem (RCPSP) is a classical scheduling problem that has received a lot of attention from the community on metaheuristics (see, e.g., [1]). In addition, a large and widely-accepted dataset is available publicly for RCPSP [14], so that algorithms can be compared on a common ground.

We propose a local search algorithm that makes use of a complex neighborhood relation based on a hybridization with a constructive heuristics. More precisely, at each iteration, we create a list of activities that are first removed altogether from the schedule, and then reinserted one by one (in their order) in the possible best position in the schedule. This neighborhood relation takes inspiration from the good results that were achieved with a similar approach for the TSP problem [3].

We perform an experimental analysis to understand the behavior of the algorithm and to tune its parameters. In addition, we compare, in a statistically-principled way, our algorithm with our implementation of a tabu search algorithm based on a combination of neighborhood relations borrowed from the literature. Finally we place our best results within the ones reported in the literature.

The outcome is that our algorithm performs favourably with the tabu search, which was reported to be among the most competitive methods. In addition, although for a definitive answer a more complete analysis is needed, we can claim that our solver is competitive with those in the literature.

T. Bartz-Beielstein et al. (Eds.): HM 2007, LNCS 4771, pp. 57–68, 2007.

The paper is organized as follows. In Section 2 we describe the RCPSP by providing the mathematical formulation, describing the public broadly-accepted dataset, and discussing related work on the problem. In Section 3 we illustrate our main local search solver and the TS-based "competitor", both in terms of search space, neighborhood relations, and specific metaheuristics. In Section 4 we report the results of our experimental analysis and we make the comparisons. Finally, in Section 5 we draw some conclusions and we discuss future work.

## 2   Resource-Constrained Project Scheduling Problem

There are two versions of the RCPSP, namely the single-mode and multi-mode one. In the multi-mode version of the problem, an activity has to be performed in one of the prescribed ways (modes) using specified amount of the resources, whereas in the single-mode version there is exactly one execution mode for each activity. In this paper we focus on the single-mode version of the problem.

### 2.1   Problem Formulation

The resource-constrained project scheduling problem (RCPSP) can be stated as follows. Given are $n$ activities $a_1, \ldots, a_n$ and $r$ renewable resources. A constant amount $R_k$ of units of resource $k$ is available at any time. Activity $a_i$ must be processed for $p_i$ time units; preemption is not allowed. During this time period a constant amount of $r_{ik}$ units of resource $k$ is occupied. All the values $R_k$, $p_i$, and $r_{ik}$ are non-negative integers.

Furthermore, there are precedence relations defined between activities. That is, we are given a directed graph $G = (V, E)$ with $V = \{1, \ldots, n\}$ such that if $(i, j) \in E$ then activity $j$ cannot start before the end of activity $i$.

The objective is to determine starting times $s_i$ for the activities $a_i$, $i = 1, \ldots, n$ in such a way that:

- at each time $t$ the total resource demand is less than or equal to the resource availability for each resource,
- the precedence constraints are fulfilled and,
- the makespan $C_{max} = \max_{i=1}^{n} c_i$, where $c_i = s_i + p_i$, is minimized.

As a generalization of the job-shop scheduling problem the RCPSP is NP-hard in the strong sense.

Let us illustrate the above definitions with the example on Fig. 1 which is taken from [22]. There are eleven activities and one single resource. The numbers associated with each node give the length of the activity and the units of resource it uses.

Clearly, a schedule is represented by the assignment of starting times of all activities. However, it can also be represented indirectly by an *activity list*, which is a permutation of all the activities.

An activity list is called feasible if it satisfies all precedence constraints, i.e. each activity has all its network predecessors as predecessors in the list.

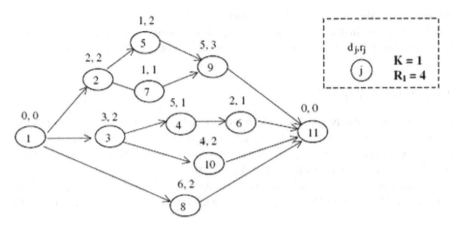

**Fig. 1.** Activity network

It is easy to see that given a feasible activity list there is a unique way to build a feasible schedule of minimal makespan, i.e., the so-called left-justified one. This is simply obtained by placing the activities in the given order one by one at the earliest possible starting time according to precedences and resource constraints. It can also be shown that there always exists an activity list that generates an optimal schedule [12].

Given the network depicted in Fig. 1, some of the (feasible) activity lists are: $\lambda = (1,2,3,5,7,9,8,4,6,10,11)$ $\alpha = (1,2,3,7,10,4,8,5,9,6,11)$ and $\beta = (1,2,3,7,4,10,8,5,6,9,11)$.

For example, the makespan of $\beta$ is 18, as the activity list $\beta$ gives rise to a schedule depicted on Fig. 2.

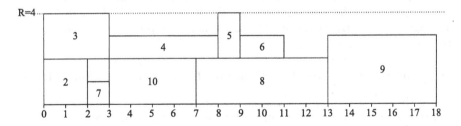

**Fig. 2.** Feasible schedule for activity list $\beta$

## 2.2 Datasets

PSPLIB [14] is a large dataset for RCPSP that is composed of 480 instances for each $n = 30, 60, 90$, and $120$, for the single-mode version of the problem. Virtually all papers dealing with RCPSP have considered this dataset as the ground for the experimental analysis.

We investigated the performance of our algorithm on various PSPLIB instances. In this work, we give preliminary results on the instances with 60 and

90 activities. We have selected ten instances for $n = 60$ and ten for $n = 90$. We selected the instances randomly, but only among the "hard" ones; that is, instances for which the known best solution is not equal to the lower bound reported.

### 2.3    Related Work

In recent years, many papers have been published on RCPSP, as reported for example in the surveys by Özdamar and Ulusoy [17] and by Brucker *et al* [4]. A great progress have been made in the solving procedures which take into account two different approaches: optimal and heuristic.

The optimal approach includes methods such as 0-1 linear programming (Mingozzi *et al* [15]) and implicit enumeration with branch and bound (Brucker *et al* [5]).

Nevertheless, the NP-hard nature of the problem makes it difficult to solve large-size projects [2], in such a way that, in practice, the use of heuristics is necessary. Therefore, besides exact algorithms many authors have developed heuristics for the RCPSP as the only feasible method of handling practical RCPSP instances (for a survey see the work by Kolisch and Hartmann [10,13]).

Among other, the one that most resembles our approach is the insertion technique [6], which is also based on insertion of activities in a schedule, but it is not based on local search and it makes use of a different representation of the search space. Furthermore this method inserts only one activity in one iteration, whereas ours inserts multiple ones in the same iteration.

## 3    Local Search for RCPSP

In order to apply the local search paradigm we need to define the search space, the cost function, the selection rule for the initial solution (Section 3.1), and the neighborhood structure (Section 3.2). Finally, we present our metaheuristic procedures (Section 3.3).

### 3.1    Search Space, Cost Function, and Initial Solution Construction

As already discussed in Section 2.1, an activity list represents a schedule. Therefore, the search space is simply given by the set of possible activity lists, i.e., permutations of the set $\{1, \ldots, n\}$. Among all activity lists only the feasible ones, i.e. those satisfying all precedence constraints, are considered as possible search states.

The cost function is simply given by the makespan of the schedule. In fact, this is the sole objective, and we do not consider schedules that violate some constraints.

The initial solution is constructed from scratch in a stepwise extension of the partial schedule (this approach is thoroughly investigated also in [12]). In each step we randomly choose one activity from the set of activities that have all predecessors already scheduled and put it at the end of the partial activity list.

## 3.2 Neighborhood Relations

We first introduce the new neighborhood relation that we call RaR (Remove and Reinsert). Then we describe the literature-based one that we use for comparison.

The main idea of RaR is to remove a fixed number $m$ of activities from the schedule and insert them back into the schedule, where $m$ is a parameter of the method.

A move $M$ is thus identified by a sequence of $m$ activities $M = \{a_{M_1}, a_{M_2}, \ldots, a_{M_m}\}$, and it is executed in a state $S$ leading to state $S'$ in the following way:

1. $S' = S - M$    // remove all activities in $M$ from $S$
2. for each $i = 1, \ldots, m$, add activity $a_{M_i}$ to $S'$ in the following way:
   - 2.1. search for the position $j$ in which the makespan of $S'$ plus $a_{M_i}$ is minimized (breaking ties randomly)
   - 2.2 insert activity $a_{M_i}$ in position $j$ in $S'$

For example, let $m = 2$ and consider the network of Fig. 1 and the state $\beta$ corresponding to the schedule of Fig. 2, and the RaR move $M = \{5, 9\}$. Starting from $\beta$, the activity list obtained in Step 1 is $\beta^{(0)} = 1, 2, 3, 7, 4, 10, 8, 6, 11$. In Step 2, we first reinsert the activity 5 in order to construct $\beta^{(1)} = 1, 2, 3, 5, 7, 4, 10, 8, 6, 11$. Activity 5 can only be reinserted after its predecessor 2, and before its successor 11 (not 9, because it was also deleted). Among the feasible positions, the one chosen gives the minimal makespan. Second, after reinsertion of 9 we get $\beta^{(2)} = 1, 2, 3, 5, 7, 9, 4, 10, 8, 6, 11$ with makespan 15 (see Fig. 3).    □

The other neighborhood relation that we use, which we call MS (for MultiShift), is actually the set-union of two neighborhoods. For the first one, the main idea is to move an activity $i$ behind some other activity $j$, which is not in precedence relation to $i$. In the literature this move is called *shift move* and it has been introduced in [1]. More formally, a move $M$ is identified by a pair $\{i, j\}$ (with $i, j \in \{1, \ldots, n\}$), such that $a_i$ is placed after $a_j$ in the schedule $L$. The move can be illustrated by

$$L = \ldots i \ldots u \ldots j \ldots \Rightarrow L' = \ldots j \, i \, u \ldots$$

where $u$ is an activity that must be scheduled after $i$. The shift of $i$ thus is accompanied by a set of other shifts so as to prevent to break precedence constraints.

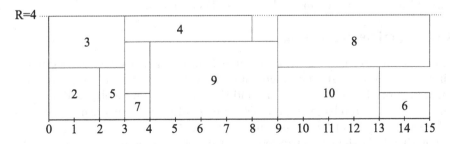

**Fig. 3.** Schedule for activity list $\beta^{(2)}$

The second component of our neighborhood works in the reversed way. Instead of moving activity $i$ after $j$, we move activity $j$ before $i$. Again, if necessary, we also make accompanying set of shifts in order to keep all precedence constraints satisfied.

These neighborhoods together are designed in such a way, that they make room for some activities to be scheduled earlier and therefore possibly shorten the makespan of the schedule.

### 3.3   Local Search Metaheuristics

The neighborhood RaR, for most of the values of $m$ that we consider, is a large neighborhood, and its exhaustive exploration turned out to be rather impractical. Therefore for this neighborhood, we resort to a simple hill-climbing strategy, based on a randomized non-ascending selection. In details, at each iteration, we draw a random move and compute its cost. If the move is improving or *sideways* it is executed, otherwise the state is unchanged. The procedure stops after a fixed number of iterations. We call this algorithm HC(RaR).

It is worth noticing that HC(RaR) can also be seen as a form of min-conflict hill-climbing (MCHC) [16], since activities are selected at random, but then are reinserted in the optimal way. In addition, like MCHC, it accepts sideways moved (for diversification).

On the other hand, for the smaller neighborhood MS, more "aggressive" metaheuristics seem to be more effective (based on preliminary experiments). Therefore, we make use of a Tabu Search [9], with a dynamic-size tabu list to implement a short term prohibition mechanism (called Robust Tabu Search in [11]). In addition, we use the standard aspiration criterion and we search for the next state by exploring the full neighborhood at each iteration. We call this algorithm TS(MS).

## 4   Experimental Analysis

In order to determine how successful our new approach is, we first explain our experimental settings (Section 4.1) and then experimental parameter tuning for both methods (Section 4.2). Next we compare both methods with best parameters for each method (Section 4.3) and finally we present comparison with methods and results reported in the literature (Section 4.4).

### 4.1   Experimental Setting

Experiments were performed on an Intel Pentium 4 (3.4 GHz) processor running Linux Suse v. 10. The algorithms have been coded in C++ exploiting the framework EASYLOCAL++ [7], and the executables were obtained using the GNU C/C++ compiler (v. 4.0.1). The statistical tests are performed using the software environment for statistical computing R [20].

For HC(RaR) the stop criterion in our experiments is based on the number of iterations, i.e. the number of feasible schedules generated. Given that the running time of a single iteration depends on $m$, in order to make a comparison fair, we normalize the number of iterations so that the total running times are approximately the same.

A typical number of iterations used in the literature (see [10,13]) is 5000. We decided to grant 5000 iterations to the largest value of $m$ and to augment the others accordingly.

In particular, for the instances with 60 activities, we have tested HC(RaR) with $m = 2, \ldots, 20$, and the running times for $m = 20$ of 5000 iteration turned out to be about 150s. All settings for both HC(RaR) and TS(MS) are made in order to run for 150s per trial. For HC(RaR), this means that the number of iterations allowed was larger for smaller $m$, reaching the highest value 23333 at $m = 2$. More precisely, we have used a heuristic formula for the number of iterations such that the runs needed approximately the same time. The formula is $max\_iter = 10500/(m + 1)(1 + \delta)$, where $\delta = (20 - m)/36$.

## 4.2   Experiments for Parameter Tuning for HC(RaR) and TS(MS)

Our first set of experiments aims at identifying the best value of $m$ for HC(RaR). Table 1 shows the results for the selected instances of size $n = 60$, for 30 trials for each instance. The table reports the average deviation w.r.t. the lower bound (obtained through the critical-path method [1]), and its standard deviation.

The outcome of the experiments show that the best results are obtained for $m = 7$ (in bold). Note that for values around 7 the results are very close and the standard deviation is relatively small for all of them.

**Table 1.** Results of HC(RaR) for $m = 2, \ldots, 20$ with time limit 150s for $n=60$

| $m$ | avg. diff. (%) | std. dev. |
|-----|----------------|-----------|
| 2   | 11.668         | 1.0497    |
| 3   | 11.364         | 0.9448    |
| 4   | 11.340         | 1.0158    |
| 5   | 11.252         | 0.9548    |
| 6   | 11.130         | 0.9713    |
| **7** | **11.107**   | 0.9636    |
| 8   | 11.122         | 0.8810    |
| 9   | 11.242         | 0.8473    |
| 10  | 11.294         | 0.9244    |
| 11  | 11.388         | 0.9131    |
| 12  | 11.583         | 0.8689    |
| 13  | 11.742         | 0.9826    |
| 14  | 11.867         | 0.9330    |
| 15  | 11.918         | 0.8784    |
| 16  | 12.152         | 0.9073    |
| 17  | 12.345         | 0.8866    |
| 18  | 12.598         | 0.9920    |
| 19  | 12.797         | 0.8610    |
| 20  | 12.880         | 0.9419    |

**Table 2.** Results of TS(MS) for different tabu lengths for $n = 60$

| Tabu length | avg. diff. (%) | std. dev. |
|:---:|:---:|:---:|
| 5 10 | 14.526 | 1.3679 |
| 5 15 | 14.473 | 1.3468 |
| 5 20 | 14.470 | 1.4237 |
| **10 15** | **14.339** | 1.3703 |
| 10 20 | 14.587 | 1.4881 |
| 10 25 | 14.364 | 1.4140 |
| 10 30 | 14.453 | 1.4528 |
| 15 20 | 14.519 | 1.5012 |
| 15 30 | 14.370 | 1.3830 |
| 20 30 | 14.485 | 1.4563 |
| 20 40 | 14.435 | 1.5786 |

**Table 3.** Results of HC(RaR) for $m = 2, \ldots, 20$ with time limit 210s for $n = 90$

| $m$ | avg. diff. (%) | std. dev. |
|:---:|:---:|:---:|
| 2 | 13.398 | 1.4224 |
| 3 | 12.842 | 1.4536 |
| 4 | 12.661 | 1.3023 |
| 5 | 12.662 | 1.4095 |
| 6 | 12.721 | 1.3402 |
| **7** | **12.654** | 1.2768 |
| 8 | 12.782 | 1.3863 |
| 9 | 13.143 | 1.2843 |
| 10 | 13.283 | 1.3608 |
| 11 | 13.333 | 1.3961 |
| 12 | 13.516 | 1.3693 |
| 13 | 13.753 | 1.5144 |
| 14 | 13.922 | 1.3413 |
| 15 | 14.061 | 1.3176 |
| 16 | 14.306 | 1.3292 |
| 17 | 14.215 | 1.2872 |
| 18 | 14.478 | 1.3614 |
| 19 | 14.654 | 1.2652 |
| 20 | 14.956 | 1.3357 |

Similarly, Table 2 shows that the best configuration for TS(MS) is with tabu length [10,15] for $n = 60$, although results for other tabu lengths are very similar.

The experiments for $n = 90$ presented in Table 3 show that the best results are obtained when $m = 7$, although results for other $m$ that are near 7 are also very competitive.

The choice of a suitable value for $m$ is clearly crucial for the behavior of any local search based on RaR. In our preliminary work [19], we conjectured that the best value of $m$ is linearly dependent on $n$, and in particular it is about $m = n/10$. In the experiments reported here, we provide a more statistically-principled comparison, and we show that the conjecture was not very precise, and, surprisingly, the best choice is the fixed value $m = 7$ for both datasets.

To test this hypothesis we did a smaller experiment on the dataset with $n = 120$, which again seems to confirm that the best choice for $m$ is 7.

### 4.3   Comparison Between HC(RaR) and TS(MS)

Table 4 shows a comparison between our two methods with problemset $n = 60$. We compared only the best performing choice of parameters for both methods, more precisely for HC(RaR) we used $m = 7$ and for TS(MS) we used tabu length [10,15]. For each instance it shows the results for 30 trials in terms of: the best value obtained, the average percentage difference w.r.t. the lower bound, and the standard deviation. The last column shows the $p$-value for the Wilcoxon test [23], which represents the confidence in the statistical difference of the two sequences of the results, in favour of HC(RaR).

The last line shows the comparison for all instances together, using the paired Wilcoxon test.

**Table 4.** Comparison between HC(RaR) and TS(MS), $n = 60$

| | | | | HC(RaR) | | | TS(MS) | | |
|---|---|---|---|---|---|---|---|---|---|
| Instance | LB | UB | best | avg. dev. (%) | std. dev. | best | avg. dev. (%) | std. dev. | p |
| j605_3 | 75 | 80 | 81 | 9.16 | 0.9568 | 82 | 12.76 | 1.8740 | 2.848e-08 |
| j605_10 | 79 | 81 | 81 | 5.11 | 1.0482 | 83 | 8.65 | 1.3437 | 3.727e-09 |
| j609_4 | 79 | 87 | 88 | 15.27 | 1.3148 | 88 | 17.38 | 1.5902 | 4.176e-05 |
| j609_5 | 77 | 85 | 88 | 14.29 | 0.8563 | 89 | 19.87 | 1.4410 | 2.928e-11 |
| j6013_7 | 80 | 87 | 88 | 12.25 | 0.9451 | 91 | 16.33 | 1.1527 | 1.156e-10 |
| j6025_5 | 86 | 98 | 98 | 16.01 | 0.8034 | 100 | 18.8 | 1.2671 | 1.095e-09 |
| j6025_7 | 83 | 90 | 91 | 11.24 | 1.1925 | 93 | 14.9 | 1.3287 | 1.531e-09 |
| j6029_5 | 102 | 110 | 112 | 11.86 | 1.2476 | 112 | 14.71 | 1.8073 | 1.274e-07 |
| j6029_8 | 96 | 103 | 104 | 9.48 | 0.7895 | 106 | 12.15 | 1.2472 | 3.401e-10 |
| j6037_8 | 88 | 93 | 93 | 6.4 | 0.4818 | 94 | 7.84 | 0.6506 | 3.916e-09 |
| ALL | – | – | – | 11.107 | 0.9636 | – | 14.339 | 1.3702 | 4.193e-06 |

Table 4 shows that HC(RaR) is clearly superior to TS(MS) in means of average deviation, namely mean of the average deviations on all the instances is 11.10% against 14.33%. Also mean of the standard deviations for HC(RaR) is significantly better than TS(MS), and the $p$ values are very close to 0, showing a confidence in the statistical difference of almost 100%.

Table 5 presents the same data for $n = 90$, and it shows similar results.

### 4.4   Comparison with Previous Results

In [13] experimental results of almost all the state-of-the-art algorithms are reported. Comparison is made with number of iterations limited to $i = 1000, 5000$ and 50000. Since we have high running times, we will compare our method with results reported for 50000 iterations.

**Table 5.** Comparison between HC(RaR) and TS(MS) for $n = 90$

| Instance | LB | UB | HC(RaR) | | | TS(MS) | | | p |
|---|---|---|---|---|---|---|---|---|---|
| | | | best | avg. dev. (%) | std. dev. | best | avg. dev. (%) | std. dev. | |
| j905_3 | 82 | 87 | 90 | 12.68 | 1.2 | 91 | 16.75 | 1.6519 | 3.493e-09 |
| j905_5 | 107 | 111 | 114 | 8.72 | 1.1926 | 118 | 13.43 | 2.714 | 6.515e-11 |
| j909_3 | 97 | 102 | 106 | 11.96 | 1.2543 | 111 | 17.8 | 1.9482 | 3.343e-11 |
| j909_7 | 102 | 109 | 113 | 13.27 | 1.2579 | 117 | 19.02 | 1.7814 | 3.673e-11 |
| j9013_3 | 104 | 108 | 113 | 10.29 | 0.8226 | 116 | 14.2 | 1.5638 | 3.076e-11 |
| j9021_6 | 95 | 106 | 108 | 17.79 | 1.8321 | 111 | 20.56 | 1.9619 | 8.481e-06 |
| j9029_5 | 114 | 121 | 125 | 11.78 | 1.0858 | 127 | 15.5 | 1.66 | 2.808e-10 |
| j9037_2 | 103 | 115 | 116 | 15.08 | 1.384 | 119 | 18.71 | 1.9137 | 1.394e-09 |
| j9037_7 | 112 | 123 | 123 | 11.85 | 1.34 | 126 | 15.98 | 2.0712 | 1.816e-10 |
| j9045_7 | 127 | 136 | 140 | 13.12 | 1.3984 | 145 | 17.11 | 1.7114 | 5.869e-11 |
| ALL | – | – | – | 12.654 | 1.2767 | – | 16.906 | 1.8977 | 8.487e-07 |

**Table 6.** Comparison between the best methods for solving RCPSP, $n = 60$

| Method | Author | avg. dev.(%) |
|---|---|---|
| Scatter search | Debels et al | 10.71 |
| GA-hybrid | Valls et al. | 10.73 |
| GA, TS-path relinking | Kochetov and Stolyar | 10.74 |
| GA-FBI | Valls et al. | 10.74 |
| GA-forw.-backw., FBI | Alcaraz et al. | 10.84 |
| **HC(RaR)** | **Pesek, Schaerf, Žerovnik** | **11.10** |
| GA-self-adapting | Hartmann | 11.21 |
| GA-activity list | Hartmann | 11.23 |
| Sampling-LFT, FBI | Tormos and Lova | 11.36 |
| TS-activity list | Nonobe and Ibaraki | 11.58 |

We should however remark that, at this stage, a fair comparison is not possible for two reasons: First, in the most relevant literature, results are reported only as the average on all 480 instances, and we haven't completed such a massive computation yet. Secondly, the computing power granted to the solvers is always expressed only in terms of visited schedules, and it is not clear how this metric applies to our solver. The most straightforward application is by granting this number of iteration to our solvers, but we have to admit that this would be too biased on our side, as our iteration is computationally much more expensive than the equivalent step in the literature (in some cases even more than 2 orders of magnitude).

Having in mind these limits of the comparison, we summarize the current results in Table 6, which shows the best ten algorithms reported (rest of the algorithms reported is in [13]). In first column shows the method used, then the author of the method and in the last column the average deviation from the critical path lower bound.

We can assert that our results are well in-line with the best solvers, also having in mind that it regards only a set of "hard" instances (which are the minority), whereas the others are averaged on all of them, but, on the other hand, we have much longer runs.

## 5  Conclusions and Future Work

We have proposed a local search solver for RCPSP based on a large neighborhood coming from a hybridization with a constructive heuristic. In our experimental analysis we have identified the best parameter setting for the solver, and compared it with our implementation, using the same technologies, of a state-of-the-art solver, namely a tabu search based on a classical neighborhood relation. The results, although preliminary, are quite encouraging.

Obviously, the first future work will be to perform a comprehensive analysis on all PSPLIB instances, so as to place the results within the relevant literature, by identifying a common ground of comparison that takes into account both results and running times.

We also plan to find new ways to improve our solver on the various levels of intervention: metaheuristic strategy, neighborhood exploration, and implementation.

Finally, in this paper we focused on the single-mode problem, whereas the multi mode problem is actually more natural and describes real life problems more accurately. Therefore, we plan to work on this other version, because we believe that our approach can be adapted to it and also provide good results.

## References

1. Baar, T., Brucker, P., Knust, S.: Tabu search algorithms and lower bounds for the resource-constrained project scheduling problem. In: Voss, S., Martello, S., Osman, I., Roucairol, C. (eds.) Meta-Heuristics: Advances and Trends in Local Search Paradigms for Optimization, pp. 1–18. Kluwer Academic Publishers, Dordrecht (1998)
2. Blazewicz, J., Lenstra, J., Kan, A.R.: Scheduling subject to resource constraints: Classification and complexity. Discrete Applied Mathematics 5, 11–24 (1983)
3. Brest, J., Žerovnik, J.: An approximation algorithm for the asymmetric traveling salesman problem. Ricerca Operativa 28, 59–67 (1999)
4. Brucker, P., Drexl, A., Möhring, R., Neumann, K., Pesch, E.: Resource-constrained project scheduling: Notation, classification, models, and methods. European Journal of Operational Research 112(1), 3–41 (1999)
5. Brucker, P., Knust, S., Schoo, A., Thiele, O.: A branch and bound algorithm for the resource-constrained project scheduling problem. European Journal of Operational Research 107(2), 272–288 (1998)
6. Christian, A., Michelon, P., Reusser, S.: Insertion techniques for static and dynamic resource-constrained project scheduling. European Journal of Operational Research 149, 249–267 (2003)
7. Di Gaspero, L., Schaerf, A.: EASYLOCAL++: An object-oriented framework for flexible design of local search algorithms. Software—Practice and Experience 33(8), 733–765 (2003)

8. Gendreau, M., Hertz, A., Laporte, G., Stan, M.: A generalized insertion heuristic for the traveling salesman problem with time windows. Operations Research 46(3), 330–335 (1998)

9. Glover, F., Laguna, M.: Tabu search. Kluwer Academic Publishers, Dordrecht (1997)

10. Hartmann, S., Kolisch, R.: Experimental evaulation of state-of-the-art heuristics for the resource-constrained project scheduling problem. European Journal of Operational Research 127(2), 394–407 (2000)

11. Hoos, H.H., Stützle, T.: Stochastic Local Search Foundations and Applications. Morgan Kaufmann Publishers, San Francisco, CA (USA) (2005)

12. Kolisch, R., Hartmann, S.: Heuristic algorithms for solving the resource-constrained project scheduling problem - classification and computational analysis. In: Weglarz, J. (ed.) Handbook on recent advances in project scheduling, pp. 147–178. Kluwer Academic Publishers, Dordrecht (1999)

13. Kolisch, R., Hartmann, S.: Experimental investigation of heuristics for resource-constrained project scheduling: An update. European Journal of Operational Research 174(1), 23–37 (2006)

14. Kolisch, R., Sprecher, A.: PSPLIB – a project scheduling library. European Journal of Operational Research 96(1), 205–216 (1997) Data available from http://129.187.106.231/psplib/

15. Mingozzi, A., Maniezzo, V., Ricciardelli, S., Bianco, L.: An exact algorithm for the resource-constrained project scheduling problem based on a new mathematical formulation. Management Science 44(5), 714–729 (1998)

16. Minton, S., Johnston, M.D., Philips, A.B., Laird, P.: Minimizing conflicts: a heuristic repair method for constraint satisfaction and scheduling problems. Artificial Intelligence 58, 161–205 (1992)

17. Özdamar, L., Ulusoy, G.: A survey on the resource-constrained project scheduling problem. IIE transactions 27(5), 574–586 (1995)

18. Palpant, M., Artigues, C., Michelon, P.: Lssper: Solving the resource-constrained project scheduling problem with large neighbourhood search. Annals of Operations Research 131(1-4), 237–257 (2004)

19. Pesek, I., Žerovnik, J.: Best insertion algorithm for resource-constrained project scheduling problem (preprint, 2006) available on http://arxiv.org/abs/0705.2137v1

20. R Development Core Team. R: A language and environment for statistical computing. R Foundation for Statistical Computing, Vienna, Austria (2005) ISBN 3-900051-07-0.

21. Solomon, M.M.: Algorithms for the vehicle routing and scheduling problems with time window constraints. Operations Research 35(2), 254–265 (1987)

22. Valls, V., Quintanilla, S., Ballestín, F.: Resource-constrained project scheduling: A critical activity reordering heuristic. European Journal of Operational Research 149, 282–301 (2003)

23. Wilcoxon, F.: Individual comparisons by ranking methods. Biometrics Bulletin 1, 80–83 (1945)

# Evolutionary Clustering Search for Flowtime Minimization in Permutation Flow Shop

Geraldo Ribeiro Filho[1], Marcelo Seido Nagano[2], and Luiz Antonio Nogueira Lorena[3]

[1] Faculdade Bandeirantes de Educação Superior
R. José Correia Gonçalves, 57
08675-130, Suzano - SP - Brazil
geraldorf@uol.com.br
[2] Escola de Engenharia de São Carlos - USP
Av. Trabalhador São-Carlense, 400
13566-590, São Carlos - SP - Brazil
drnagano@usp.br
[3] Instituto Nacional de Pesquisas Espaciais - INPE/LAC
Av. dos Astronautas, 1758
12227-010, São José dos Campos - SP - Brazil
lorena@lac.inpe.br

**Abstract.** This paper deals with the Permutation Flow Shop scheduling problem with the objective of minimizing total flow time, and therefore reducing in-process inventory. A new hybrid metaheuristic Genetic Algorithm - Cluster Search is proposed for the scheduling problem solution. The performance of the proposed method is evaluated and results are compared with the best reported in the literature. Experimental tests show the new method superiority for the test problems set, regarding the solution quality.

## 1 Introduction

This paper deals with Permutation Flow Shop scheduling problems, which consists of finding a sequence for the jobs that optimises some schedule performance measure. Usually, such measures are the maximum completion time (makespan), and the total flowtime. As it is well known, the first measure is associated with an efficient utilization of resources, and the second one with a faster response to job processing, therefore reducing in-process inventory. In this paper we introduce a hybrid meta-heuristic method with the objective of minimizing the total flowtime.

This production scheduling problem is NP-complete [1,2], therefore the search for an optimal solution is of more theoretical than practical importance.

In the last ten years a number of heuristic methods have been introduced with the objective of minimizing total flowtime, or equivalently the mean flowtime in permutation flow shops. These heuristic methods can be divided into two

T. Bartz-Beielstein et al. (Eds.): HM 2007, LNCS 4771, pp. 69–81, 2007.

main classes: construction methods and improvement methods. The literature on construction methods includes the heuristics proposed by Ahmadi and Bagchi [3], Rajendran and Chaudhuri [4], Rajendran [5], Ho [6], Wang et al. [7], Woo and Yim [8], Liu and Reeves [9], Allahverdi and Aldowaisan [10], Framinan and Leisten [11], Framinan et al. [12], Li et al. [13] and Nagano and Moccellin [14].

Improvement methods such as ant-colony optimization algorithm proposed by Rajendran and Ziegler [15] and swarm optimization algorithm proposed by Tasgetiren et al. [16], start from an initial permutation, which is usually generated by a construction method, and then iteratively generate a sequence of improved permutations. It is obvious that improvement methods generate significantly better solutions than construction ones.

Rajendran and Ziegler [15] introduce two heuristics. The first algorithm extends the ideas of the ant-colony algorithm by Stuetzle [17], called max-min ant system (MMAS), by incorporating the summation rule suggested by Merkle and Middendorf [18] and a new proposed local search technique. The second ant-colony algorithm is newly developed. These ant-colony algorithms were applied to 90 benchmark problems taken from Taillard [19]. Considering the minimization of makespan the comparison shows that the two proposed ant-colony algorithms perform better, on an average, than the MMAS. Subsequently, by considering the objective of minimizing the total flowtime of jobs, a comparison of solutions yielded by the proposed ant-colony algorithms with the best heuristic solutions known for the benchmark problems, as published in an extensive study by Liu and Reeves [9], is carried out. The comparison shows that the proposed ant-colony algorithms are clearly superior to the heuristics analyzed by Liu and Reeves. For 83 out of 90 problems considered, better solutions have been found by the two proposed ant-colony algorithms, as compared to the solutions reported by Liu and Reeves [9].

Like many optimization problems, scheduling are commonly approached by evolutionary techniques. Cotta and Fernandez [20] applied memetic algorithms to planning, scheduling and timetabling, and Kleeman and Lamont [21] have studied multi-objective evolutionary algorithms (MOEA) with fixed and variable chromosome length applied to the flow-shop and job-shop problems.

Recently Tasgetiren et al. [16] presented a particle swarm optimization algorithm (PSO) to solve the permutation flow shop sequencing problem with the objectives of minimizing makespan and the total flowtime of jobs. For this purpose, a heuristic rule called the smallest position value (SPV) borrowed from the random key representation of Bean [22] was developed to enable the continuous particle swarm optimization algorithm to be applied to all classes of sequencing problems. In addition, a very efficient local search, called variable neighborhood search (VNS), was embedded in the PSO algorithm to solve the well known benchmark suites in the literature. The PSO algorithm was applied to both the 90 benchmark instances provided by Taillard [19], and the 14,000 random, narrow random and structured benchmark instances provided by Watson et al. [23]. For makespan criterion, the solution quality was evaluated according to the best known solutions provided either by Taillard [19], or Watson et al. [23]. The total flowtime criterion was evaluated with the best known solutions provided by Liu and Reeves [9], and Rajendran and Ziegler [15]. For the total flowtime criterion, 57 out of the 90 best known solutions reported by Liu and Reeves [9], and Rajendran and Ziegler [15] were improved whereas for the makespan criterion, 195 out of the 800 best known solutions for the random and narrow random problems reported by Watson et al. [23] were improved by the VNS version of the PSO algorithm.

Based on the literature examination we have made, the aforementioned metaheuristic PSO-VNS presented by Tasgetiren et al. [16] yields the best solutions for total flowtime minimization in a permutation flow shop.

## 2 Clustering Search

The metaheuristic Clustering Search (CS), proposed by Oliveira and Lorena [24,25], consists of a solution clustering process to detect supposedly promising regions in the search space. The objective of the detection of these regions as soon as possible is to adapt the search strategy. A region can be seen as a search subspace defined by a neighborhood relation.

The CS has an iterative clustering process, simultaneously executed with a heuristic, and tries to identify solution clusters that deserve special interest. The regions defined by these clusters must be explored, as soon as they are detected, by problem specific local search procedures. The expected result of more rational use of local search is convergence improvement associated with reduction of computational effort.

CS tries to locate promising regions by using clusters to represent these regions. A cluster is defined by a triple $G = (C, r, \beta)$ where $C$, $r$ and $\beta$ are, respectively, the center, the radius of a search region around the center, and a search strategy associated with the cluster.

The center $C$ is a solution that represents the cluster, identify its location in the search space, and can be changed along the iterative process. Initially the centers can be obtained randomly, and progressively tend to move to more promising points in the search space. The radius $r$ defines the maximum distance from the center to consider a solution being inside the cluster. For example, the radius $r$ could be defined as the number moves to change a solution into another. The CS admits a solution to be inside of more than one cluster. The strategy $\beta$ is a procedure to intensify the search, in which existing solutions interact with each other to create new ones.

The CS consists of four components, conceptually independent, with different attributions: a metaheuristic (ME), an iterative clustering process (IC), a cluster analyzer (CA), and a local optimization algorithm (LO). Figure 1 shows a representation of the four components, the search space and the clusters centers.

The ME component works as a full time iterative solution generator. The algorithm is independently executed from the other CS components, and must be able to continuously generate solutions for the clustering process. Simultaneously, the clusters are maintained as containers for these solutions. This process works as a loop in which solutions are generated along the iterations.

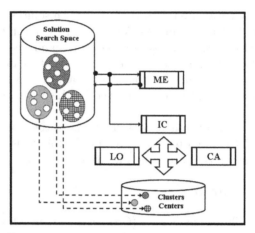

**Fig. 1.** Clustering Search Conceptual Diagram

The objective of the IC component is to associate similar solutions to form a cluster, keeping a representative one of them as the cluster center. The IC is implemented as an online process where the clusters are feed with the solutions produced by the ME. A maximum number of clusters is previously defined to avoid unlimited cluster generation. A distance metric must be defined also previously to evaluate solutions similarity for the clustering process. Each solution received by IC is inserted into the cluster having the center most similar to it, causing a perturbation in this center. This perturbation is called assimilation and consists of the center update according to the inserted solution.

The CA component provides an analysis of each cluster, at regular time intervals, indicating probable promising clusters. The so called cluster density $\lambda_i$ measures the $i$-th cluster activity. For simplicity, $\lambda_i$ can be the cluster's number of assimilated solutions. When $\lambda_i$ reach some threshold, meaning that ME has produced a predominant information model, the cluster must be more intensively investigated to accelerate its convergence to better search space regions. CA is also responsible for the removal of low density clusters, allowing new and better clusters to be created, while preserving the most active clusters. The clusters removal does not interfere with the set of solutions being produced by ME, as they are kept in a separate structure.

Finally, the LO component is a local search module that provides more intensive exploration of promising regions represented by active clusters. This process runs after CA has determined a highly active cluster. The local search corresponds to the $\beta$ element that defines the cluster and is a problem specific local search procedure.

## 3   Evolutionary Clustering for the Permutation Flow Shop Problem

This research has used a metaheuristic called Evolutionary Clustering Search (ECS) proposed by Oliveira and Lorena [24,25] that combines Genetic

Algorithms (GA) and Custering Search, and has applied it to the Permutation Flow Shop problem. The ECS uses a GA to implement the ME component of the CS and generate solutions that allow the exploration of promising regions by the other components of CS. A pseudo-code representation of the ECS is shown in Figure 2.

```
Procedure ECS-FS()
Begin
  Initialize population P;
  Initialize clusters set C;
  While (stop condition == false) do Begin
    While (i < new_individuals) do Begin
      parent1 = Selected from best 40% of P;
      parent2 = Selected from the whole P;
      offspring = Crossover(parent1, parent2);
      Local_Search_LS1(offspring) with 60% probability;
      If (Insert_into_P(offspring))
        Assimilate_or_create_cluster(offspring, C);
      i = i + 1;
    End;
    For each cluster c in C
      If (High_assimilation(c))
        Local_Search_LS2(c);
  End;
End;
```

**Fig. 2.** Pseudo-code for the ECS algorithm

As ECS has presented good performance in previous applications, and considering the accelerated convergence provided by CS when compared with pure, non hybrid, algorithms, the aim of this work was to attempt to beat the best results recently produced and found in the literature, even with larger computer times, characteristic of evolutionary processes. Seeking originality, this was another reason to apply CS in this research.

The Evolutionary Clustering Search for Flow Shop (ECS-FS) presented in this work has some modifications from the original CS general concept presented in the previous section.

As the quality of the individuals in the initial population is important for the GA performance, to ensure this quality, the population initialization was done with a variation of the method known as NEH, presented by Nawaz et al. [26]. The original form of NEH initially sort a set of $n$ tasks according to non-descending values of the sum of task processing times by all machines. The two first tasks in the sorted sequence are scheduled to minimize the partial flow time. The remaining tasks are then sequentially inserted into the schedule in the position that minimizes the partial flow time.

The chromosome representation used in the GA was a $n$ element vector, one element for each task, storing the position of that task in the solution schedule. After several tests, the population size was fixed in 500 individuals to make room for good individuals produced by NEH and its variation, together with randomly generated individuals to provide diversity.

The very fist individual inserted into the initial population was generated by the NEH procedure. Part of the other individuals was generated by a variation of the NEH in which the two tasks from the sorted sequence to be first scheduled were randomly chosen from the whole sequence. The rest of the sequence was then scheduled the same way as the original NEH.

To ensure some degree of diversity in the initial population, the maximum number of individuals generated by the modified NEH was given by

$$\min\left(\frac{n*(n-1)}{4}, \frac{500}{2}\right) , \tag{1}$$

and the remain part of the initial population was filled with randomly generated schedules.

The evaluation of the population individuals was made by the minimization of the total flow time. The individual insertion routine kept the population sorted, and the best individual, the one with the lowest total flow time, occupied the first position in the population. The insertion routine was also responsible for maintain only one copy of each individual in the population.

A cluster set initialization process was created to take advantage of the good individuals in the GA initial population. This routine scanned the population, from the best individual to the worst, creating new clusters or assimilating the individuals into clusters already created. A new cluster was created when the distance from the individual to the center of any cluster was larger than $r = 0.85*n$, and the individual was used to represent the center of new cluster. Otherwise, the individual was assimilated by the cluster with the closest center. The distance measure from an individual to the cluster center was taken as the number of swaps necessary to transform the individual into the cluster center. Starting from the very first, each element of the individual was compared to its equivalent in the cluster center, at the same position. When non coincident elements were found the rest of the individual chromosome was scanned to find the same element found in the cluster center, and make a swap. At the end, the individual was identical to the cluster center, and the number of swaps was considered as a distance measure. The clusters initialization process ended when the whole population was scanned or when a maximum of 200 clusters were created. Both the cluster radius and the maximum number of clusters are parameters which values were chosen after several tests, with the objective to work with all problem classes used for tests.

The assimilation of an individual by a cluster was based on the Path Relinking procedure presented by Glover [27]. Starting from the individual chromosome, successive swaps were made until the chromosome became identical to the cluster center. The pair of genes chosen to swap was the one that more reduced, or less increased, the chromosome total flow time. At each swap the new chromosome configuration was evaluated. At the end of the transformation, the cluster center was moved to (replaced by) the individual, or the intermediary chromosome, that has the best evaluation better than the current center. If no such improvement was possible, the cluster center remains the same.

At each iteration of the GA, 50 new individuals were created and possibly inserted into the population. The stop condition used was the maximum of 100 iterations or 20 consecutive iterations with no new individuals being inserted, as the population could have one single copy of each individual.

The new individual generation was made by randomly selecting two parents, one from the best 40% of the population, called the base parent, and the other from the entire population, called the guide parent. A crossover process known as Block Order Crossove (BOX), presented by Syswerda [28], was then applied to both parents, generating a single offspring by copying blocks of genes from the parents. In this work the offspring was generated with 50% of its genes coming from each parent. Several other recombination operators are studied and empirically evaluated by Cotta and Troya [29]. Investigation regarding position-oriented recombination operators are also possible in further studies. Figure 3 illustrates the BOX crossover.

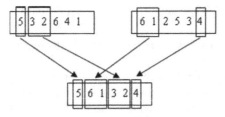

**Fig. 3.** BOX Crossover

The number of new individuals created at each iteration, the stop condition, the part of the population from which comes the base parent for crossover, and contribution of each parent in the crossover process are all parameters which values were obtained after several tests.

After the crossover, the offspring had a probability of 60% to be improved by a local search procedure called LS1, shown in Figure 4.

This procedure used two neighborhood types: permutation and insertion. The permutation neighborhood around an individual was obtained by swapping every possible pair of chromosome genes, producing $n*(n-1)/2$ different individuals. The insertion neighborhood was obtained by removing every gene from its position, and inserting it in each other position in the chromosome, producing $n*(n-1)$ different individuals.

The new individual was then inserted into the population in the position relative to its evaluation, shifting ahead the subsequent part of the population, and therefore removing the last, and worst, individual.

The successfully inserted individuals were then processed by the IC component of ECS-FS. This procedure tried to find the cluster having the closest center and of which radius $r$ the individual was within. When such cluster could found, the individual was assimilated, otherwise a new cluster was created having the individual as its center. New clusters were created only if the ECS-FS had not reached the 200 clusters limit. Tests have shown that the number of cluster tends

```
Procedure LS1(current_solution)
Begin
  cs = current_solution;
  stop = false;
  While (stop == false) do Begin
    P = Permutation_neighborhood(cs);
    sp = First s in P that eval(s) < eval(cs), or eval(s) < eval(t) for all t in P;
    I = Insertion_neighborhood(cs);
    si = First s in I that eval(s) < eval(cs), or eval(s) < eval(t) for all t in I;
    If (eval(sp) < min(eval(si), eval(cs))) then
      cs = sp;
    else
      If (eval(si) < min(eval(sp), eval(cs))) then
        cs = si;
      else
        stop = true;
  End;
  Return cs;
End
```

**Fig. 4.** Pseudo-code for the LS1 Local Search Procedure

to increase at very first ECS iterations, and slowly decrease as iterations continue and the ECS removes the less active clusters.

After the generation of new individuals, its improvement and insertion into the population, the ECS-FS executed its CA component. This cluster analysis procedure performed two tasks: remove the clusters that had no assimilations in the last 5 iterations, and take every cluster that had any assimilation in the current iteration and ran it through a local optimization procedure, called LS2 and shown in Figure 5, corresponding to the LO component of ECS-FS.

Again, the probability with which an offspring ran trough local search before being inserted into the population and the number of iterations without assimilation used to delete clusters was parameters which values were chosen after several tests.

Along the ECS-FS processing the best cluster was kept saved. At the end of the ECS-FS execution, the center of the best cluster found so far was taken as the final solution produced by the algorithm.

## 4   Computational Experiments

The performance evaluation of the proposed hybrid heuristic method ECS-FS, was made through computational experiments using the Taillard [19] test problems. These problem are divided into $n$ tasks and $m$ machines sets, each set having ten instances. Results were compared with those reported in the works of Liu and Reeves [9], Rajendran and Ziegler [15], Li et al. [13] and Tasgetiren et al. [16].

For this work, the ECS-FS code was written in the C programming language and was executed on a Pentium IV, 3.0 GHz, 1 GByte RAM personal computer.

Two statistic measures were used to performance evaluation: the success rate and the average relative deviation. The first is given by the ratio between the

```
Procedure LS2(current_solution)
Begin
  cs = current_solution;
  stop = false;
  While (stop == false) do Begin
    I = Insertion_neighborhood(cs);
    si = First s in I that eval(s) < eval(cs), or eval(s) < eval(t) for all t in I;
    If (eval(si) < eval(cs)) then Begin
      cs = si;
      P = Permutation_neighborhood(cs);
      sp = First s in P that eval(s) < eval(cs), or eval(s) < eval(t) for all t in P;
      If (eval(sp) < eval(cs)) then
        cs = sp;
    End else Begin
      Pnh = Permutation_neighborhood(cs);
      sp = Scan Pnh until sp is better than cs, or sp is the best in Pnh;
      If (eval(sp) < eval(cs))
        cs = sp;
      else
        stop = true;
    End;
  End;
  Return cs;
End
```

**Fig. 5.** Pseudo-code for the LS2 Local Search Procedure

number of problems for which a method produced the minimum total flow time given by all compared methods, and the number of problems solved in a problem set. The second shows the deviation obtained by a method $h$ from the minimum total flow time as above, and is given by

$$RD_h = \left( \frac{F_h - F_*}{F_*} \right) \qquad (2)$$

where $F_h$ is the total flow time given by the method $h$, and $F_*$ is the minimum total flow time given by all compared methods, for a given test problem.

## 5  Analysis of Results

The ECS-FS performance was evaluated comparing its results with the two ant colony based algorithms (M-MMAS and PACO) shown by Rajendran and Ziegler [15], and the particle swarm method (PSO) shown by Tasgetiren et al. [16].

Nine test problems classes were considered, each one having ten instances. The classes were defined according to the number of tasks $n$ being equal to 20, 50 and 100, and for each one of these values, the number of machines $m$ being equal to 5, 10 and 20.

Table 1 presents the best solution obtained by the methods for each instance and shows the superiority of ECS-FS to the others for the test problems. The listed results of ECS-FS are the best out of 10 repeats. For a total of 90 instances the ECS-FS produced equal or better solutions for 82 of them, corresponding

**Table 1.** New Best Known solution for Taillard's benchmarks for Flowtime minimisation in Permutation Flow Shop

| n m | M − MMAS | PACO | PSO$_{vns}$ | ECS − FS | n m | M − MMAS | PACO | PSO$_{vns}$ | ECS − FS |
|---|---|---|---|---|---|---|---|---|---|
| 20 5 | 14056 | 14056 | **14033** | **14033** | 50 20 | 127348 | 126962 | 128622 | **126315** |
| | 15151 | 15214 | **15151** | **15151** | | 121208 | 121098 | 122173 | **119502** |
| | 13416 | 13403 | **13301** | **13301** | | 118051 | 117524 | 118719 | **116910** |
| | 15486 | 15505 | **15447** | **15447** | | 123061 | 122807 | 123028 | **121031** |
| | **13529** | **13529** | 13529 | 13529 | | 119920 | 119221 | 121202 | **118914** |
| | 13139 | **13123** | **13123** | **13123** | | 122369 | 122262 | 123217 | **121087** |
| | 13559 | 13674 | **13548** | **13548** | | 125609 | 125351 | 125586 | **123340** |
| | 13968 | 14042 | **13948** | **13948** | | 124543 | 124374 | 125714 | **123005** |
| | 14317 | 14383 | **14295** | **14295** | | 124059 | 123646 | 124932 | **122203** |
| | 12968 | 13021 | **12943** | **12943** | | 126582 | 125767 | 126311 | **124785** |
| 20 10 | 20980 | 20958 | **20911** | **20911** | 100 5 | 257025 | 257886 | **254762** | 254911 |
| | **22440** | 22591 | **22440** | **22440** | | 246612 | 246326 | 245315 | **243943** |
| | **19833** | 19968 | **19833** | **19833** | | 240537 | 241271 | 239777 | **239002** |
| | 18724 | 18769 | **18710** | **18710** | | 230480 | 230376 | **228872** | 228888 |
| | 18644 | 18749 | **18641** | **18641** | | 243013 | 243457 | 242245 | **241659** |
| | **19245** | **19245** | 19249 | **19245** | | 236225 | 236409 | **234082** | 234172 |
| | 18376 | 18377 | **18363** | **18363** | | 243935 | 243854 | 242122 | **241753** |
| | **20241** | 20377 | **20241** | **20241** | | 234813 | 234579 | 232755 | **232315** |
| | **20330** | **20330** | **20330** | **20330** | | 252384 | 253325 | 249959 | **249608** |
| | **21320** | 21323 | **21320** | **21320** | | 246261 | 246750 | 244275 | **244210** |
| 20 20 | **33623** | **33623** | 34975 | **33623** | 100 10 | 305004 | 305376 | 303142 | **301176** |
| | 31604 | 31597 | 32659 | **31587** | | 279094 | 278921 | 277109 | **276902** |
| | **33920** | 34130 | 34594 | **33920** | | 297177 | 294239 | 292465 | **290844** |
| | 31698 | 31753 | 32716 | **31661** | | 306994 | 306739 | 304676 | **304377** |
| | 34593 | 34642 | 35455 | **34557** | | 290493 | 289676 | 288242 | **287545** |
| | 32637 | 32594 | 33530 | **32564** | | 276449 | 275932 | 272790 | **272635** |
| | 33038 | **32922** | 33733 | **32922** | | 286545 | 284846 | 282440 | **282381** |
| | 32444 | 32533 | 33008 | **32412** | | 297454 | 297400 | **293572** | 294119 |
| | 33625 | 33623 | 34446 | **33600** | | 309664 | 307043 | 305605 | **304964** |
| | 32317 | 32317 | 33281 | **32262** | | 296869 | 297182 | 295173 | **294362** |
| 50 5 | 65768 | 65546 | 65058 | **64838** | 100 20 | 373756 | 372630 | 374351 | **371391** |
| | 66828 | 68485 | 68298 | **68159** | | 383614 | 381124 | 379792 | **376383** |
| | 64166 | 64149 | 63577 | **63453** | | 380112 | 379135 | 378174 | **374599** |
| | 69113 | 69359 | 68571 | **68310** | | 380201 | 380765 | 380899 | **378550** |
| | 70331 | 70154 | 69698 | **69477** | | 377268 | 379064 | 376187 | **374426** |
| | 67563 | 67664 | 67138 | **66902** | | 381510 | 380464 | 379248 | **377567** |
| | 67014 | 66600 | **66338** | 66355 | | 381963 | 382015 | 380912 | **378367** |
| | 64863 | 65123 | 64638 | **64471** | | 393617 | 393075 | 392315 | **389680** |
| | 63735 | 63483 | 63227 | **63068** | | 385478 | 380359 | 382212 | **380152** |
| | 70256 | 69831 | 69195 | **69092** | | 387948 | 388060 | 386013 | **383928** |
| 50 10 | 89599 | 88942 | 88031 | **87683** | | | | | |
| | 83612 | 84549 | 83624 | **83535** | | | | | |
| | 81655 | 81338 | 80609 | **80365** | | | | | |
| | 87924 | 88014 | 87053 | **86934** | | | | | |
| | 88826 | 87801 | 87263 | **86865** | | | | | |
| | 88394 | 88269 | 87255 | **86969** | | | | | |
| | 90686 | 89984 | **89259** | 89304 | | | | | |
| | 88595 | 88281 | **87192** | 87316 | | | | | |
| | 86975 | 86995 | **86102** | 86213 | | | | | |
| | 89470 | 89238 | 88631 | **88534** | | | | | |
| 50 20 | 127348 | 126962 | 128622 | **126315** | | | | | |
| | 121208 | 121098 | 122173 | **119502** | | | | | |
| | 118051 | 117524 | 118719 | **116910** | | | | | |
| | 123061 | 122807 | 123028 | **121031** | | | | | |
| | 119920 | 119221 | 121202 | **118914** | | | | | |
| | 122369 | 122262 | 123217 | **121087** | | | | | |
| | 125609 | 125351 | 125586 | **123340** | | | | | |
| | 124543 | 124374 | 125714 | **123005** | | | | | |
| | 124059 | 123646 | 124932 | **122203** | | | | | |
| | 126582 | 125767 | 126311 | **124785** | | | | | |

to 91.1%, and 59 solutions (65.5%) were inedited, better than the previous best found in the literaure.

Table 2 presents success rates for problems classes and shows that the ECS-FS had its success rate varying from 70% to 100%. This table shows also that the difference from the ECS-FS average relative deviation to the other method deviation is emphasized, reinforcing the proposed method superiority.

The quality of the initial population individuals, allied to diversity, and the performance of the local search routines, can be considered key factors for the quality of the final solutions.

Average processing time had a large variation from 8.62 seconds for the smallest problems class, with 20 tasks and 5 machines, to 7 hours and 39 minutes for the largest problems class used in this work, with 100 tasks and 20 machines.

**Table 2.** Success Rate (a) and Average Relative Deviation (b)

| $n$ | $m$ | $M - MMAS$ | $PACO$ | $PSO_{uns}$ | $ECS - FS$ |
|---|---|---|---|---|---|
| 20 | 5 | $20^a$ | 20 | **100** | **100** |
|    |   | $0.1975^b$ | 0.4544 | **0.0000** | **0.0000** |
| 20 | 10 | 60 | 20 | 90 | **100** |
|    |   | 0.0492 | 0.3235 | 0.0021 | **0.0000** |
| 20 | 20 | 20 | 20 | 0 | **100** |
|    |   | 0.1195 | 0.1892 | 2.8278 | **0.0000** |
| 50 | 5 | 0 | 0 | 10 | **90** |
|    |   | 1.1302 | 0.9450 | 0.2452 | **0.0026** |
| 50 | 10 | 0 | 0 | 30 | **70** |
|    |   | 1.4196 | 1.1569 | 0.1841 | **0.0322** |
| 50 | 20 | 0 | 0 | 0 | **100** |
|    |   | 1.2852 | 0.9780 | 1.8421 | **0.0000** |
| 100 | 5 | 0 | 0 | 30 | **70** |
|    |   | 0.8733 | 0.9921 | 0.1638 | **0.0104** |
| 100 | 10 | 0 | 0 | 10 | **90** |
|    |   | 1.2714 | 0.9834 | 0.2189 | **0.0186** |
| 100 | 20 | 0 | 0 | 0 | **100** |
|    |   | 1.0678 | 0.8361 | 0.6627 | **0.0000** |

## 6   Conclusion

The main objective of this work was apply CS to the Permutation Flow Shop Scheduling Problem of original and inedited form. Experimental results presented in the tables have shown that the ECS-FS method had superior performance regarding success rate and average relative deviation, compared with the best results found in the literature for the considered Flow Shop test problems, using the in-process inventory reduction, or minimization of the total flow time, as the performance measure. The computational effort was acceptable for practical applications.

The classic optimization problem of task schedule in Flow Shop has been the object of intense research in the last 50 years. For practical applications this problem may be considered already solved, although, because of its complexity it still remains as a target for the search for heuristic and metaheuristic methods with better efficiency and solution quality, taking into account that the problem is NP-hard.

The research related in this paper was motivated by the above considerations, and have tried to rescue the essential characteristics of metaheuristic methods, balance between solution quality and computational efficiency, simplicity and implementation easiness.

# References

1. Garey, M.R., Johnson, D.S., Sethi, R.: The Complexity of flowshop and jobshop scheduling. Mathematics of Operations Research 1, 117–129 (1976)
2. Rinnooy Kan, A.H.G.: Machine Scheduling Problems: Classification, Complexity, and Computations. Nijhoff, The Hahue (1976)
3. Ahmadi, R.H., Bagchi, U.: Improved lower bounds for minimizing the sum of completion times of n jobs over m machines. European Journal of Operational Research 44, 331–336 (1990)
4. Rajendran, C., Chaudhuri, D.: An efficient heuristic approach to the scheduling of jobs in a flowshop. European Journal of Operational Research 61, 318–325 (1991)
5. Rajendran, C.: Heuristic algorithm for scheduling in a flowshop to minimise total flowtime. International Journal Production Economics 29, 65–73 (1993)
6. Ho, J.C.: Flowshop sequencing with mean flowtime objective. European Journal of Operational Research 81, 571–578 (1995)
7. Wang, C., Chu, C., Proth, J.M.: Heuristic approaches for n/m/F/ Ci scheduling problems. European Journal of Operational Research 96, 636–644 (1997)
8. Woo, D.S., Yim, H.S.: A heuristic algorithm for mean flowtime objective in flowshop scheduling. Computers and Operations Research 25, 175–182 (1998)
9. Liu, J., Reeves, C.R.: Constructive and composite heuristic solutions to the $P//\sum C_i$ scheduling problem. European Journal of Operational Research 132, 439–452 (2001)
10. Allahverdi, A., Aldowaisan, T.: New heuristics to minimize total completion time in m-machine flowshops. International Journal of Production Economics 77, 71–83 (2002)
11. Framinan, J.M., Leisten, R.: An efficient constructive heuristic for flowtime minimisation in permutation flow shop. OMEGA 31, 311–317 (2003)
12. Framinan, J.M., Leisten, R., Ruiz-Usano, R.: Comparison of heuristics for flowtime minimisation in permutation flowshops. Computer and Operations Research 32, 1237–1254 (2005)
13. Li, X., Wang, Q., Wu, C.: Efficient composite heuristics for total flowtime minimization in permutation flow shops. Omega (2006) doi: 10.1016-j.omega.2006.11.003
14. Nagano, M.S., Moccellin, J.V.: Reducing mean flow time in permutation flow shop. Journal of the Operational Research Society (2007) doi: 10.1057/palgrave.jors.2602395
15. Rajendran, C., Ziegler, H.: Ant-colony algorithms for permutation flowshop scheduling to minimize makespan/total flowtime of jobs. European Journal of Operational Research 155, 426–438 (2004)
16. Tasgetiren, M.F., Liang, Y.C., Sevkli, M., Gencyilmaz, G.: A particle swarm optimization algorithm for makespan and total flowtime minimization in the permutation flowshop sequencing problem. European Journal of Operational Research 177, 1930–1947 (2007)
17. Stutzle, T.: Applying iterated local search to the permutation flowshop problem. Technical Report, AIDA-98-04, Darmstad University of Technology, Computer Science Department, Intelletics Group, Darmstad, Germany (1998)

18. Merkle, D., Middendorf, M.: An ant algorithm with a new pheromone evaluation rule for total tardiness problems. In: Oates, M.J., Lanzi, P.L., Li, Y., Cagnoni, S., Corne, D.W., Fogarty, T.C., Poli, R., Smith, G.D. (eds.) EvoWorkshops 2000. LNCS, vol. 1803, pp. 287–296. Springer, Heidelberg (2000)
19. Taillard, E.: Benchmarks for basic scheduling problems. European Journal of Operational Research 64, 278–285 (1993)
20. Cotta, C., Fernandez, A.J.: Memetic Algorithms in Planning, Scheduling and Timetabling. Evolutionary Schedulling, pp. 1–30. Springer, Heidelberg (2006)
21. Kleeman, M.P., Lamont, G.B.: Scheduling of flow-shop, job-shop, and combined scheduling problems using MOEAs with fixed and variable length chromosomes. Evolutionary Schedulling, pp. 49–100. Springer, Heidelberg (2006)
22. Bean, J.C.: Genetic algorithm and random keys for sequencing and optimization. ORSA Journal on Computing 6, 154–160 (1994)
23. Watson, J.P., Barbulescu, L., Whitley, L.D., Howe, A.E.: Contrasting structured and random permutation flowshop scheduling problems: Search space topology and algorithm performance. ORSA Journal of Computing 14, 98–123 (2002)
24. Oliveira, A.C.M., Lorena, L.A.N.: Detecting promising areas by evolutionary clustering search. In: Bazzan, A.L.C., Labidi, S. (eds.) Advances in Artificial Intelligence. LNCS (LNAI), pp. 385–394. Springer, Heidelberg (2004)
25. Oliveira, A.C.M., Lorena, L.A.N.: Hybrid evolutionary algorithms and clustering search. In: Grosan, C., Abraham, A., Ishibuchi, H. (eds.) Hybrid Evolutionary Systems - Studies in Computational Intelligence. SCI Series, vol. 75, pp. 81–102. Springer, Heidelberg (2007)
26. Nawaz, M., Enscore, E.E., Ham, I.: A heuristic algorithm for the m-machine, n-job flow-shop sequencing problem. OMEGA 11, 91–95 (1983)
27. Glover, F.: Tabu search and adaptive memory programing: Advances, applications and challenges. In: Interfaces in Computer Science and Operations Research, pp. 1–75. Kluwer Academic Publishers, Dordrecht (1996)
28. Syswerda, G.: Uniform crossover in genetic algorithms. In: International Conference on Genetic Algorithms (ICGA), Virginia, USA, pp. 2–9 (1989)
29. Cotta, C., Troya, J.M.: Genetic Forma Recombination in Permutation Flowshop Problems. Evolutionary Computation 6, 25–44 (1998)

# A Hybrid ILS Heuristic to the Referee Assignment Problem with an Embedded MIP Strategy

Alexandre R. Duarte[1], Celso C. Ribeiro[2],
and Sebastián Urrutia[3]

[1] Department of Computer Science, Catholic University of Rio de Janeiro,
Rua Marquês de São Vicente 225, Rio de Janeiro, RJ 22453-900, Brazil
[2] Department of Computer Science, Universidade Federal Fluminense,
Rua Passo da Pátria 156, Niterói, RJ 24210-240, Brazil
[3] Department of Computer Science, Universidade Federal de Minas Gerais,
Av. Antônio Carlos 6627, Belo Horizonte, MG 31270-010, Brazil
{aduarte,celso}@inf.puc-rio.br, surrutia@dcc.ufmg.br

**Abstract.** Optimization in sports is a field of increasing interest. A novel problem in sports management is the Referee Assignment Problem, in which a limited number of referees with different qualifications and availabilities should be assigned to a set of games already scheduled. We extend and improve a previous three-phase approach for this problem, based on a constructive heuristic, a repair heuristic to make the initial solutions feasible, and an ILS improvement heuristic. We propose a new constructive algorithm based on a greedy criterion to build initial solutions. Furthermore, we develop a hybridization strategy in which a mixed integer programming exact algorithm replaces the original neighborhood-based local search within the ILS heuristic. Computational experiments are performed for large realistic instances. The use of time-to-target-solution-value plots is emphasized in the evaluation of the numerical results, illustrating the efficiency and the robustness of the new approach. The proposed hybridization of MIP with local search can be extended to other metaheuristics and applications, opening a new research avenue to more robust algorithms.

## 1 Introduction

Optimization in sports is a field of increasing interest and different optimization techniques have been applied to solve problems arising from sports scheduling and management, see e.g. Ribeiro and Urrutia [27], Easton et al. [8], and Rasmussen and Trick [22] for some state-of-the-art reviews. Playoff elimination [28], the traveling tournament problem [2,9,29], and the scheduling of a college basketball conference [21] are some examples of optimization problems in sports.

T. Bartz-Beielstein et al. (Eds.): HM 2007, LNCS 4771, pp. 82–95, 2007.

A novel problem in sports management is the Referee Assignment Problem (RAP) [6,7]. Sport games are regulated by rules that depend on the sport and tournament. The officiating crew is a group of referees that is responsible to ensure that all rules are respected in a game. The number of referees compounding a crew may vary, depending on the sport, league, and tournament: soccer games usually require three referees, while basketball games require two. Each member of an officiating crew fills a refereeing position in a game. For example, in a regular soccer game, there are one head umpire and two side judges, totalizing three refereeing positions to be filled with referees. In some applications, managers make pre-assignments to satisfy some specific requirements. The referee assignment problem consists in assigning referees to the empty refereeing positions (not yet assigned) for all games of a league or tournament.

This problem appears in regional amateur leagues in the United States. Amateur leagues of several sports, such as baseball, basketball and soccer, have hundreds of games every weekend in different divisions. As an example, in the MOSA (Monmouth & Ocean Counties Soccer Association) league, New Jersey, boys and girls of ages 8 to 18 make up six divisions per age and gender group with six teams per division, totalizing 396 games every Sunday. In a single league in California there might be up to 500 soccer games in a weekend, to be refereed by hundreds of certified referees.

Referee assignment is subject to a number of criteria. Games in higher divisions may require higher-skilled referees. Due to the shortage of certified referees, each of them may officiate several games in the same day. Some referees may have to travel between facilities and the traveling times have to be considered. Some players or their relatives may also act as referees and a natural constraint is that a referee cannot officiate a game in which he/she or a relative is scheduled to play. In some applications, each referee declares the target number of games he/she is willing to officiate and the objective consists in minimizing the sum over all referees of the absolute value of the difference between the target and the actual number of games assigned to each referee. In case the referees are able to officiate in different facilities, some objectives may be the minimization of the total number of inter-facility travels, the minimization of the total traveling time, or the minimization of the idle times between consecutive games assigned to the same referee. Tournament organizers may also want referee assignments matching preferences regarding the facilities, divisions, and time slots where the referees officiate.

Referee assignment problems in other contexts have been addressed in [4,10] [11,30]. In the next section, we state the variant of the referee assignment problem considered in this work, whose decision version is NP-complete [7]. Section 3 describes some improvements to the three-phase solution strategy proposed in [7] (a constructive procedure, a repair heuristic to make solutions feasible, and an ILS improvement heuristic): a new constructive heuristic and a mixed-integer programming (MIP) local search strategy. Numerical results illustrating the improvements observed with these extensions are presented in Section 4. Concluding remarks are drawn in the last section.

## 2   Problem Statement

We consider the general problem, in which each game has a number of refereeing positions to be assigned to referees. The games are previously scheduled and the facilities and time slots for each game are known beforehand. In our approach, referees are assigned to empty refereeing positions, not to games. This allows not only to handle referee assignment problems in sports requiring different number of referees, but also in tournaments where different games of the same sport may need different numbers of referees due to the game division or importance. Games with pre-assigned referees to some refereeing positions can also be handled by this approach. Each refereeing position to be filled by a referee is called a refereeing slot.

Let $S = \{1, \ldots, n\}$ be the set of refereeing slots. Each refereeing slot $j \in S$ has to be filled by a referee with a minimum skill level $q_j$, which is previously determined and often related to the tournament division. Usually, a division corresponds to a set of teams formed by players under a certain age and with the same gender, e.g. boys under 16 years old. Let $R = \{1, \ldots, m\}$ be the set of referees, represented by their indices. Each referee $i \in R$ has a certain skill level, denoted by $p_i$, defining the refereeing slots in which he/she can officiate. Referees may declare their unavailability to officiate at certain time slots. Furthermore, each referee $i \in R$ establishes $M_i$ as the maximum number of games he/she is able to officiate and $T_i$ as the target number of games he/she is willing to officiate. Travels are not allowed, i.e. referees that officiate more than one game in the same day must be assigned to games that take place at the same facility.

Several objectives can be optimized. We consider the minimization of the sum over all referees of the absolute value of the difference between the target and the actual number of games assigned to each referee. Briefly, the referee assignment problem consists in assigning referees to all refereeing slots associated to games scheduled to a given time interval (typically, a day or a weekend), minimizing the objective function described above and satisfying the set of hard constraints listed below:

(a) all refereeing slots must be filled for all games;
(b) referees cannot officiate more than one game in overlapping time slots;
(c) referees cannot officiate games in time slots where they are unavailable;
(d) referees must meet the minimum skill level established for each refereeing slot;
(e) referees cannot officiate more than a given maximum number of games; and
(f) each referee can officiate in only one facility.

## 3   Solution Approach and Extensions

This work proposes some extensions to the three-phase heuristic approach proposed by Duarte et al. [7] to solve the referee assignment problem. This approach consists in a constructive procedure, a repair heuristic to make solutions feasible, and an improvement heuristic to improve feasible solutions. The constructive

```
1  Algorithm RefereeAssignmentHeuristic
2  Solution ← BuildGreedySolution();
3  if Solution is not feasible then
4  |   Solution ← RepairHeuristic(Solution);
5  end
6  if Solution is feasible then
7  |   Solution ← ImprovementHeuristic(Solution);
8  |   return Solution;
9  else
10 |   return no feasible solution was found;
11 end
```

**Algorithm 1.** Referee assignment heuristic

heuristic emphasizes feasibility. The repair heuristic is used when the constructive algorithm fails to build a feasible solution. The improvement heuristic follows the paradigm of the Iterated Local Search (ILS) metaheuristic [19,20]. The main steps of the pseudo-code of three-phase heuristic for referee assignment are given in Algorithm 1.

The pseudo-code of the ILS improvement heuristic is presented in Algorithm 2. It starts by a first improving local search applied to the initial feasible solution. Since the local search involves moves that change referee assignments for only one facility at a time, it should be applied to every facility. Next, a perturbation involving one pair of facilities is applied to the current solution. Each perturbation is followed by two applications of the local search, once to each of the facilities of the pair involved in the perturbation. The solution obtained after local search is accepted if it satisfies some acceptance criterion. A new perturbation is applied and the above steps are repeated, until an stopping criterion based on the maximum number of iterations is met.

```
1  Algorithm ImprovementHeuristic(Solution)
2  foreach facility f do
3  |   Solution ← LocalSearch(f, Solution);
4  end
5  for i = 1, ..., MaxIterations do
6  |   NewSolution ← Perturbation(Solution);
7  |   Let f_1 and f_2 be the facilities involved in the perturbation;
8  |   NewSolution ← LocalSearch(f_1, NewSolution);
9  |   NewSolution ← LocalSearch(f_2, NewSolution);
10 |   Solution ← AcceptanceCriterion(Solution, NewSolution);
11 end
12 return Solution;
```

**Algorithm 2.** ILS improvement heuristic

We present next two extensions to improve the three-phase algorithm and to make it more efficient and robust. Section 3.1 proposes a new constructive algorithm that privileges not only feasibility, but also assignments leading to low-cost solutions. Section 3.2 presents a MIP local search strategy to replace the neighborhood-based procedure in Algorithm 2. The code of the algorithm presented in [7] was also optimized and its parameters appropriately tuned.

### 3.1   Improved Greedy Constructive Heuristic

The constructive heuristic originally proposed in [7] was designed to build feasible initial solutions, regardless of their value. Solutions built by this heuristic may have several referees officiating too many games, while others officiate just a few or even none.

To overcome this weakness, we propose a new constructive heuristic with the same structure of the latter, but more focused into finding better initial solutions, even at the cost of violating some constraints. In this new version, a first assignment pass limits the number of refereeing slots assigned to each referee at its target number of games. Only in a second pass, after all referees have been handled and if there are still unassigned slots, the heuristic assigns additional games to each referee, respecting the maximum number of games for each of them.

The main steps of the pseudo-code of this improved greedy heuristic are presented in Algorithm 3. We denote by $S^u$ the set of all unassigned refereeing slots, by $R^{HF}$ the set of referees associated with a hard facility constraint, and by $R^{NHF}$ the set of referees with no hard facility constraint, i.e. $R = R^{HF} \cup R^{NHF}$. These sets are initialized respectively in lines 2, 3, and 4. The loop in lines 5 to 12 is performed until all referees associated with hard facility constraints have been examined and assigned to as many refereeing slots as possible. The number of refereeing slots assigned to each referee in this phase is limited to his/her target number of games. Next, the loop in lines 13 to 21 attempts to fill the remaining unassigned refereeing slots with referees without hard facility constraints. Again, the algorithm limits the number of refereeing slots assigned to each referee to his/her target number of games. A greedy criterion is applied in line 15 to select a facility $f$ with the strongest need for referees with a certain skill level $\bar{p}$ computed in line 14. The computation of the greedy criterion is based on two measures: (a) an estimate of the minimum number of referees with skill level $\bar{p}$ needed to officiate at facility $f$ and (b) the number of unassigned refereeing slots in facility $f$ with minimum skill level less than or equal to $\bar{p}$.

If unassigned refereeing slots still remain at line 22, the loop in lines 23 to 28 attempts to fill the remaining unassigned refereeing slots with referees that are currently officiating games at the same facility where these slots take place. At this point, the algorithm limits the number of refereeing slots assigned to each referee to his/her maximum number of games. Finally, if there are still unassigned refereeing slots at line 30, then the loop in lines 31 to 35 makes infeasible assignments to complete the solution.

Once again, we stress the importance of quick and effective procedures for finding initial solutions for hard combinatorial problems in sports, as already noticed by Ribeiro and Urrutia [29].

---

**1  Algorithm** *BuildImprovedGreedySolution*()

**2**  $S^u \leftarrow \{j = 1, \ldots, n : \sum_{i=1}^{m} x_{ij} = 0\}$;

**3**  $R^{HF} \leftarrow \{i = 1, \ldots, m :$  referee i plays at least one game$\}$;

**4**  $R^{NHF} \leftarrow R - R^{HF}$;

**5  while** $R^{HF} \neq \emptyset$ **do**

**6**  $\quad$ Randomly select a referee $i \in R^{HF}$;

**7**  $\quad$ $R^{HF} \leftarrow R^{HF} - \{i\}$;

**8**  $\quad$ Let $f$ be the facility where referee $i$ plays a game;

**9**  $\quad$ **forall** $j \in S^u :$ refereeing slot $j$ takes place at facility $f$ **do**

**10**  $\quad\quad$ **if** $\sum_{j=1}^{n} x_{ij} < T_i$ and referee $i$ can be assigned to refereeing slot $j$
$\quad\quad$ **then** set $x_{ij} \leftarrow 1$ and $S^u \leftarrow S^u - \{j\}$;

**11**  $\quad$ **end**

**12  end**

**13  while** $S^u \neq \emptyset$ and $R^{NHF} \neq \emptyset$ **do**

**14**  $\quad$ $\overline{p} \leftarrow \max_{i \in R^{NHF}} \{p_i\}$;

**15**  $\quad$ Let $f$ be the facility with the strongest need for referees with skill level equal to $\overline{p}$;

**16**  $\quad$ Randomly select a referee $i \in R^{NHF}$ with $p_i = \overline{p}$;

**17**  $\quad$ $R^{NHF} \leftarrow R^{NHF} - \{i\}$;

**18**  $\quad$ **forall** $j \in S^u :$ refereeing slot $j$ takes place at facility $f$ **do**

**19**  $\quad\quad$ **if** $\sum_{j=1}^{n} x_{ij} < T_i$ and referee $i$ can be assigned to refereeing slot $j$
$\quad\quad$ **then** set $x_{ij} \leftarrow 1$ and $S^u \leftarrow S^u - \{j\}$;

**20**  $\quad$ **end**

**21  end**

**22  if** $S^u \neq \emptyset$ **then**

**23**  $\quad$ **forall** $j \in S^u$ **do**

**24**  $\quad\quad$ Let $f$ be the facility where $j$ takes place;

**25**  $\quad\quad$ **forall** $i \in Q(f)$ **do**

**26**  $\quad\quad\quad$ **if** $\sum_{i=1}^{m} x_{ij} = 0$ and referee $i$ can be assigned to refereeing slot $j$ **then** set $x_{ij} \leftarrow 1$ and $S^u \leftarrow S^u - \{j\}$;

**27**  $\quad\quad$ **end**

**28**  $\quad$ **end**

**29  end**

**30  if** $S^u \neq \emptyset$ **then**

**31**  $\quad$ **forall** $j \in S^u$ **do**

**32**  $\quad\quad$ Let $f$ be the facility where refereeing slot $j$ takes place;

**33**  $\quad\quad$ Randomly select a referee $i \in Q(f)$;

**34**  $\quad\quad$ Set $x_{ij} \leftarrow 1$ and $S^u \leftarrow S^u - \{j\}$;

**35**  $\quad$ **end**

**36  end**

**37  return** Solution : $\{(i,j) : i \in R, j \in S$ and $x_{ij} = 1\}$;

---

**Algorithm 3.** Improved greedy constructive heuristic

## 3.2   MIP Local Search Strategy

The local search procedure performed within Algorithm [2] considers two types of moves:

- swap moves: referees assigned to two refereeing slots are swapped (such moves do not change the number of games assigned to each referee), and
- replace moves: the referee assigned to a refereeing slot is replaced by another referee (such moves increase by one the number of games assigned to one referee and decrease by one the number of games assigned to the other).

As referees cannot be assigned to games at different facilities, only moves involving referees that officiate at the same facility (or do not officiate at all) are allowed. In the first phase of the local search procedure, only improving moves are accepted. The second phase also accepts moves leading to solutions at least as good as the current one, using a list of forbidden moves to prevent cycles. The latter is separated in two parts: first, only replace moves are considered; next, only swap moves.

Given the facility $\ell$ where the referees involved in the move officiate, the moves applied to the current solution only change the assignments involving referees and refereeing slots at this facility. Briefly, the local search solves a smaller instance of the referee assignment problem, optimizing the assignments of the referees that officiate at facility $\ell$. Therefore, we propose to substitute the local search applied after each perturbation by the exact solution of the integer programming model (1)-(7) associated with both facilities $\ell_1$ and $\ell_2$ involved in perturbations performed within Algorithm 2:

$$\text{minimize} \sum_{i \in Q(\ell_1 \cup \ell_2)} d_i \tag{1}$$

subject to:

$$d_i = \left| T_i - \sum_{j \in P(\ell_1 \cup \ell_2)} x_{ij} \right| \quad \forall i \in Q(\ell_1 \cup \ell_2) \tag{2}$$

$$\sum_{i \in Q(\ell_1 \cup \ell_2)} x_{ij} = 1 \quad \forall j \in P(\ell_1 \cup \ell_2) \tag{3}$$

$$\sum_{j \in P(\ell_1 \cup \ell_2)} x_{ij} \leq M_i \quad \forall i \in Q(\ell_1 \cup \ell_2) \tag{4}$$

$$x_{ij} + x_{ij'} \leq 1 \quad \forall i \in Q(\ell_1 \cup \ell_2), \quad \forall j \in P(\ell_1 \cup \ell_2), \quad \forall j' \in C(j) \tag{5}$$

$$\sum_{j \in U(i)} x_{ij} = 0 \quad \forall i \in Q(\ell_1 \cup \ell_2) \tag{6}$$

$$x_{ij} \in \{0, 1\} \quad \forall i \in Q(\ell_1 \cup \ell_2), \quad \forall j \in P(\ell_1 \cup \ell_2), \tag{7}$$

where:

$$x_{ij} = \begin{cases} 1, & \text{if referee } i \in Q(\ell_1 \cup \ell_2) \text{ is assigned to slot } j \in P(\ell_1 \cup \ell_2) \\ 0, & \text{otherwise.} \end{cases}$$

In the model above, $P(\ell_1 \cup \ell_2) \subseteq S$ denotes the set of refereeing slots associated to games that take place at facility $\ell_1$ or at facility $\ell_2$ in the current solution, $Q(\ell_1 \cup \ell_2) \subseteq R$ denotes the set of referees currently selected to officiate at facility $\ell_1$ or at facility $\ell_2$, $C(j) \subseteq P(\ell_1 \cup \ell_2)$ is the set of refereeing slots conflicting with slot $j \in P(\ell_1 \cup \ell_2)$ (i.e. refereeing slots overlapping with $j$), and $U(i) \subseteq P(\ell_1 \cup \ell_2)$ represents the set of refereeing slots to which referee $i \in Q(\ell_1 \cup \ell_2)$ cannot be assigned due to a lower skill level or to his/her unavailability.

The objective function (1) states that the sum over all referees officiating at facilities $\ell_1$ or $\ell_2$ of the slack between their target and actual numbers of scheduled games is minimized. Constraints (2) enforce that the slack $d_i$ is equal to the absolute value of the difference between the target and actual numbers of games assigned to referee $i \in Q(\ell_1 \cup \ell_2)$. Constraints (3) ensure that every refereeing slot of games taking place at facilities $\ell_1$ or $\ell_2$ must be assigned to exactly one referee. Constraints (4) establish the upper bound to the number of refereeing slots that can be assigned to each referee. Constraints (5) ensure that refereeing slots with timetabling conflicts cannot be assigned to the same referee. Constraints (6) prevent assignments that violate minimum skill level and unavailability restrictions. Constraints (7) establish the integrality of the decision variables.

# 4   Computational Experiments

The computational experiments reported in this section aim to evaluate the impact of each of the proposed improvements and to compare the new heuristic with that in [7]. They were performed on an AMD Athlon 1800 processor with 768 Mbytes of RAM memory running Windows 2000[TM]. All codes were implemented in C. The implementation details and parameter settings have been described in detail in [7]. Version 9.1 of the commercial integer programming solver CPLEX was used to implement the exact MIP local search strategy reported in Section 3.2.

We report results for five algorithm versions evaluated in the experiments: (1) the original heuristic (3phase), (2) the improved and code optimized version of the latter (opt3phase), (3) the previous with the greedy constructive heuristic (opt3phase+greedy), (4) the code optimized version with the MIP local search (opt3phase+MIP), (5) the complete version including the greedy constructive algorithm and the MIP local search (opt3phase+greedy+MIP). They were applied to all randomly generated instances proposed in [7] and available from [5]. These instances have up to 500 games and 1000 referees.

We selected three instances with 500 games and 750 referees, different numbers of facilities (65 and 85), and different patterns for the target number of games to illustrate the behavior of the algorithms.

The first experiment compares the original heuristic 3phase with its improved version opt3phase. Table 1 summarizes average numerical results over ten runs of each algorithm for each instance. The fourth column displays the time $T_1$ (in seconds) given to each algorithm in two different situations: 30 and 120 seconds.

**Table 1.** Improved heuristic (500 games and 750 referees)

| Facilities | Pattern | Heuristic | $T_1$ (s) | Value | $T_2$ (s) |
|---|---|---|---|---|---|
| 65 | $P_0$ | 3phase | 30.0 | 600.20 | 25.52 |
| | | opt3phase | 30.0 | 582.80 | 26.83 |
| | | 3phase | 120.0 | 577.80 | 99.39 |
| | | opt3phase | 120.0 | 569.80 | 99.33 |
| 65 | $P_1$ | 3phase | 30.0 | 998.20 | 28.21 |
| | | opt3phase | 30.0 | 965.20 | 27.96 |
| | | 3phase | 120.0 | 966.00 | 109.45 |
| | | opt3phase | 120.0 | 946.40 | 104.95 |
| 85 | $P_0$ | 3phase | 30.0 | 596.80 | 27.89 |
| | | opt3phase | 30.0 | 565.40 | 28.68 |
| | | 3phase | 120.0 | 544.60 | 109.47 |
| | | opt3phase | 120.0 | 534.60 | 97.68 |

The next column shows the best solution value obtained by each algorithm within this time limit. The sixth column displays the computation time $T_2$ (in seconds) each heuristic took to find the best solution found. The code optimized heuristic obtained better solutions in all cases.

We also used time-to-target (TTT) plots to evaluate and compare the algorithm versions and the proposed extensions. These plots display on the ordinate axis the probability that an algorithm will find a solution at least as good as a given target value within a given running time, shown on the abscissa axis. TTT plots were used by Feo, Resende, and Smith [12] and have been advocated by Hoos and Stützle [16,17,18] as a way to characterize the running times of stochastic algorithms for combinatorial optimization. They have been used in a number of computational studies, see also e.g. [1,3,13,15,23,24,25,26].

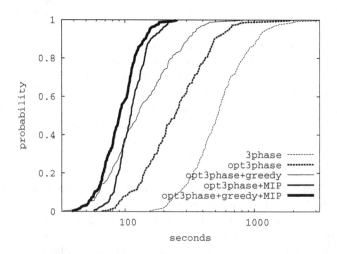

**Fig. 1.** Instance with 65 facilities and pattern $P_0$ (hard target: 562)

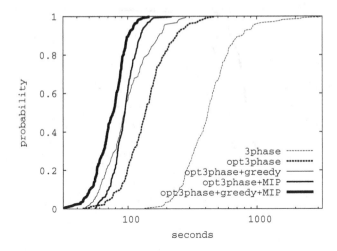

**Fig. 2.** Instance with 85 facilities and pattern $P_0$ (hard target: 529)

We discuss how TTT plots are generated, following closely Aiex, Resende, and Ribeiro [1]. These plots show the empirical distributions of the random variable *time-to-target solution value* for different algorithms, instances, and target values. To plot the empirical distribution, we fix a solution target value and run each algorithm 200 times, recording the running time when a solution with cost at least as good as the target value is found. For each algorithm, we associate with the $i$-th sorted running time $t_i$ a probability $p_i = (i - \frac{1}{2})/200$ and plot the points $z_i = (t_i, p_i)$, for $i = 1, \ldots, 200$. We notice that the most to the left a curve appears in a TTT plot, the better the corresponding algorithm is (since it takes less time to find the target value for any given probability). The runs were interrupted after 60 minutes of computation time without finding a solution as good as the target value.

Figures 1 to 3 depict some illustrative results for the five algorithm versions on the same three selected test instances. Hard target solution values have been considered in these experiments. The numerical results show that both the improved greedy constructive heuristic (**opt3phase+greedy**) and the MIP local search strategy (**opt3phase+MIP**) improved the code optimized (**opt3phase**) and the original (**3phase**) versions. The complete version (**opt3phase+greedy+MIP**) with the two extensions was faster and more robust than the others. We notice that the curves associated to the variants **opt3phase** and **3phase** are not complete in Figure 3, since many runs did not reach the target solution value within the 60 minutes time limit. The contributions of the greedy construction and of the exact MIP local search are clear.

The next experiment illustrates the contribution of each component of the improved 3-phase solution approach. Considering the instance with 65 facilities and pattern $P_0$, the first bar in Figure 4 shows the solution value after the application of the new greedy constructive heuristic. The second illustrates the solution

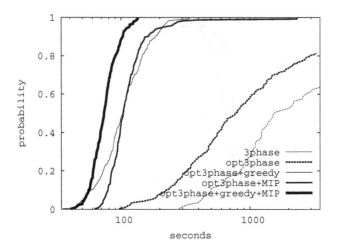

**Fig. 3.** Instance with 65 facilities and pattern $P_1$ (hard target: 926)

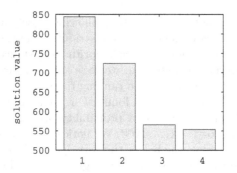

**Fig. 4.** Components of the heuristic

value after the MIP local search applied to the solution obtained by the new greedy constructive heuristic. The third bar displays the solution value obtained by the ILS heuristic running for 15,000 iterations (approximately 1000 seconds

**Table 2.** 500 games, 750 referees, 65 facilities, and pattern $P_0$

| Instance | Feasible | MIP LS | ILS iteration | ILS (1000 s) |
|---|---|---|---|---|
| $I_1$ | 853.00 | 736.80 | 732.40 | 549.00 |
| $I_2$ | 893.20 | 767.60 | 763.00 | 507.00 |
| $I_3$ | 856.00 | 748.60 | 739.60 | 544.60 |
| $I_4$ | 886.20 | 784.00 | 778.00 | 534.60 |
| $I_5$ | 893.60 | 774.80 | 770.00 | 552.80 |
| Average | 876.40 | 762.36 | 756.60 | 537.60 |
| Normalized | 1.00 | 0.87 | 0.86 | 0.61 |

**Table 3.** 500 games, 750 referees, 65 facilities, and pattern $P_1$

| Instance | Feasible | MIP LS | ILS iteration | ILS (1000 s) |
|---|---|---|---|---|
| $I_1$ | 1193.00 | 1124.20 | 1119.40 | 903.60 |
| $I_2$ | 1062.00 | 983.20 | 977.20 | 724.40 |
| $I_3$ | 1126.40 | 1036.00 | 1031.40 | 809.80 |
| $I_4$ | 1208.80 | 1123.20 | 1118.20 | 923.60 |
| $I_5$ | 1144.80 | 1053.40 | 1049.40 | 803.40 |
| Average | 1147.00 | 1064.00 | 1059.12 | 832.96 |
| Normalized | 1.00 | 0.93 | 0.92 | 0.73 |

**Table 4.** 500 games, 750 referees, 85 facilities, and pattern $P_0$

| Instance | Feasible | MIP LS | ILS iteration | ILS (1000 s) |
|---|---|---|---|---|
| $I_1$ | 864.20 | 760.00 | 755.60 | 508.00 |
| $I_2$ | 973.80 | 875.80 | 872.00 | 638.40 |
| $I_3$ | 954.60 | 852.20 | 846.40 | 588.60 |
| $I_4$ | 953.60 | 844.60 | 840.40 | 613.00 |
| $I_5$ | 942.80 | 820.80 | 817.40 | 537.00 |
| Average | 937.80 | 830.68 | 826.36 | 577.00 |
| Normalized | 1.000 | 0.89 | 0.88 | 0.62 |

of running time). The last result corresponds to the solution value obtained by the ILS heuristic running for 10,800 seconds. This plot further illustrates that each component of the heuristic plays a relevant role in the quality of the final solution.

To further illustrate the results displayed in Figure 4, we give in Tables 2 to 4 the numerical results obtained by the full heuristic on the same instances considered in [7]. Each row reports average results over ten runs of heuristic opt3phase+greedy+MIP: the average solution value of the first feasible solution found, the average solution value after the first application of the MIP local search, the average solution value after the first complete iteration of the ILS improvement phase (perturbation followed by MIP local search), and the average solution value obtained by the ILS improvement phase after 1000 seconds of computation time. Average and normalized results over the five instances are also presented in each table, showing the contribution of each phase of the complete heuristic to the final solution found.

## 5 Concluding Remarks

We presented a promising hybrid approach to embed MIP strategies within metaheuristics. The local search phase of an ILS heuristic is replaced by an exact procedure, following the same lines proposed by Fischetti and Lodi [14]. This hybridization allowed to find hard target solution values in smaller processing times. The extension of this approach to other metaheuristics is straightforward.

These results open a new research avenue, showing that the hybridization of metaheuristics with exact algorithms may lead to faster and more robust algorithms.

We also illustrated the importance of a quick and effective construction procedure to build initial solutions using a greedy criterion. We are currently working on some extensions of the referee assignment problem by addressing further constraints of real-life applications, such as the existence of hard and soft links between some referees. Decision makers may also want referee assignments matching preferences regarding the facilities, divisions, and time slots where the referees officiate.

# References

1. Aiex, R.M., Resende, M.G.C., Ribeiro, C.C.: Probability distribution of solution time in GRASP: An experimental investigation. Journal of Heuristics 8, 343–373 (2002)
2. Anagnostopoulos, A., Michel, L., Van Hentenryck, P., Vergados, Y.: A simulated annealing approach to the traveling tournament problem. Journal of Scheduling 9, 177–193 (2006)
3. Buriol, L.S., Resende, M.G.C., Ribeiro, C.C., Thorup, M.: A hybrid genetic algorithm for the weight setting problem in OSPF/IS-IS routing. Networks 46, 36–56 (2005)
4. Dinitz, J.H., Stinson, D.R.: On assigning referees to tournament schedules. Bulletin of the Institute of Combinatorics and its Applications 44, 22–28 (2005)
5. Duarte, A.R.: Challenge referee assignment problem instances, last visited on (May 23, 2007) Online document at http://www.esportemax.org/rapopt
6. Duarte, A.R., Ribeiro, C.C., Urrutia, S.: Referee assignment in sports tournaments. In: Proceedings of the 6th International Conference on the Practice and Theory of Automated Timetabling, Brno, pp. 394–397 (2006)
7. Duarte, A.R., Ribeiro, C.C., Urrutia, S., Haeusler, E.H.: Referee assignment in sport leagues. Lecture Notes in Computer Science (to appear)
8. Easton, K., Nemhauser, G.L., Trick, M.: Sports scheduling. In: Leung, J.T. (ed.) Handbook of Scheduling: Algorithms, Models and Performance Analysis, pp. 52.1–52.19. CRC Press, Boca Raton, USA (2004)
9. Easton, K., Nemhauser, G.L., Trick, M.A.: The traveling tournament problem: Description and benchmarks. In: Walsh, T. (ed.) CP 2001. LNCS, vol. 2239, pp. 580–585. Springer, Heidelberg (2001)
10. Evans, J.R.: A microcomputer-based decision support system for scheduling umpires in the American baseball league. Interfaces 18, 42–51 (1988)
11. Evans, J.R., Hebert, J.E., Deckro, R.F.: Play ball - The scheduling of sports officials. Perspectives in Computing 4, 18–29 (1984)
12. Feo, T.A., Resende, M.G.C., Smith, S.H.: A greedy randomized adaptive search procedure for maximum independent set. Operations Research 42, 860–878 (1994)
13. Festa, P., Pardalos, P.M., Resende, M.G.C., Ribeiro, C.C.: Randomized heuristics for the MAX-CUT problem. Optimization Methods and Software 7, 1033–1058 (2002)
14. Fischetti, M., Lodi, A.: MIP models for MIP heuristics. In: Talk presented at Matheuristics 2006: 1st Workshop on Mathematical Contributions to Metaheuristics, Bertinoro (2006)

15. Gent, I.P., Hoos, H.H., Rowley, A.G.D., Smyth, K.: Using stochastic local search to solve quantified Boolean formulae. In: Rossi, F. (ed.) CP 2003. LNCS, vol. 2833, pp. 348–362. Springer, Heidelberg (2003)
16. Hoos, H.H., Stützle, T.: Evaluating Las Vegas algorithms - Pitfalls and remedies. In: Proceedings of the 14th Annual Conference on Uncertainty in Artificial Intelligence, pp. 238–245 (1998)
17. Hoos, H.H., Stützle, T.: Towards a characterisation of the behaviour of stochastic local search algorithms for SAT. Artificial Intelligence 112, 213–232 (1999)
18. Hoos, H.H., Stützle, T.: Local search algorithms for SAT: An empirical evaluation. Journal of Automated Reasoning 24, 421–481 (2000)
19. Lourenço, H.P., Martin, O., Stützle, T.: Iterated Local Search. In: Glover, F., Kochenberger, G. (eds.) Handbook of Metaheuristics, pp. 321–353. Kluwer Academic Publishers, Dordrecht (2002)
20. Martin, O., Otto, S.W.: Combining simulated annealing with local search heuristics. Annals of Operations Research 63, 57–75 (1996)
21. Nemhauser, G.L., Trick, M.A.: Scheduling a major college basketball conference. Operations Research 46, 1–8 (1997)
22. Rasmussen, R.V., Trick, M.A.: Round robin scheduling - A survey. Technical report, Department of Operations Research, University of Aarhus (2006)
23. Resende, M.G.C., Ribeiro, C.C.: A GRASP with path-relinking for private virtual circuit routing. Networks 41, 104–114 (2003)
24. Resende, M.G.C., Ribeiro, C.C.: Greedy randomized adaptive search procedures. In: Glover, F., Kochenberger, G. (eds.) Handbook of Metaheuristics, pp. 219–249. Kluwer Academic Publishers, Dordrecht (2003)
25. Resende, M.G.C., Ribeiro, C.C.: Parallel Greedy Randomized Adaptive Search Procedures. In: Alba, E. (ed.) Parallel Metaheuristics: A new class of algorithms, pp. 315–346. Wiley, Chichester (2005)
26. Ribeiro, C.C., Rosseti, I.: Efficient parallel cooperative implementations of GRASP heuristics. Parallel Computing 33, 21–35 (2007)
27. Ribeiro, C.C., Urrutia, S.: OR on the ball: Applications in sports scheduling and management. OR/MS Today 31, 50–54 (2004)
28. Ribeiro, C.C., Urrutia, S.: An application of integer programming to playoff elimination in football championships. International Transactions in Operational Research 12, 375–386 (2005)
29. Ribeiro, C.C., Urrutia, S.: Heuristics for the mirrored traveling tournament problem. European Journal of Operational Research 179, 775–787 (2007)
30. Wright, M.B.: Scheduling English cricket umpires. Journal of the Operational Research Society 42, 447–452 (1991)

# On the Combination of Constraint Programming and Stochastic Search: The Sudoku Case

Rhydian Lewis

Pryfysgol Caerdydd/Cardiff University,
Cardiff Business School,
Colum Drive, Cardiff,
Wales
lewisR9@cf.ac.uk

**Abstract.** Sudoku is a notorious logic-based puzzle that is popular with puzzle enthusiasts the world over. From a computational perspective, Sudoku is also a problem that belongs to the set of NP-complete problems, implying that we cannot hope to find a polynomially bounded algorithm for solving the problem in general. Considering this feature, in this paper we demonstrate how a metaheuristic-based method for solving Sudoku puzzles (which was reported by the same author in an earlier paper), can actually be significantly improved if it is coupled with Constraint Programming techniques. Our results, which have been gained through a large amount of empirical work, suggest that this combination of techniques results in a hybrid algorithm that is significantly more powerful than either of its constituent parts.

## 1 Introduction

Sudoku is a popular puzzle that appears regularly in a variety of newspapers, books, and puzzle magazines worldwide. Although originating in the United States in the late 1970s, it was actually in Japan in the 1980s that the puzzle gained mainstream popularity. It was also here where it was given the name "Sudoku", which can be loosely translated in English as "solitary number".

In its simplest form, Sudoku can be defined as follows. Given an $n^2 \times n^2$ grid divided into $n^2$ distinct $n \times n$ *boxes* (denoted by the bold lines in fig. 1), the aim is to fill the grid so that three separate criteria are met:

1. Each row of cells contains the integers 1 through to $n^2$ exactly once;
2. Each column of cells contains the integers 1 through to $n^2$ exactly once;
3. Each $n \times n$ box contains the integers 1 through to $n^2$ exactly once.

In this paper we will refer to the value of $n$ as the *order* of a puzzle.

Typically some of the cells in a Sudoku grid will have been pre-filled by the puzzle master (see fig. 1). The player will then use these to logically determine the values for other cells in the grid, eventually allowing him-or-her to complete the puzzle. As can be imagined, *how many* and *which* cells the puzzle-master chooses to fill will therefore be particularly important if the puzzle is

T. Bartz-Beielstein et al. (Eds.): HM 2007, LNCS 4771, pp. 96–107, 2007.

| | | | | | | | | 8 |
|---|---|---|---|---|---|---|---|---|
| 1 | 8 | | | | 2 | 3 | | |
| | 6 | | | 5 | 7 | | | 1 |
| | 7 | | 9 | 6 | | | | |
| | 9 | | 7 | | 4 | | 1 | |
| | | | 8 | 1 | | 4 | | |
| 6 | | | 2 | 4 | | | 8 | |
| | | 4 | 5 | | | | 9 | 3 |
| 5 | | | | | | | | |

**Fig. 1.** Example of an order-3 Sudoku puzzle. This particular grid is logic-solvable

to be enjoyable for the player. Generally speaking, a "good" puzzle (from the player's perspective) should be configured in such a way so that is *logic-solvable* – that is, the player should be able to complete the puzzle in a logical sequence of steps using forward-chaining logic only (obviously the deductive abilities of different players will vary). In particular, a player should not usually be required to make random choices, especially when the grid is still quite empty, because if this guess turns out to be wrong, he-or-she will then have to go through the unsatisfying process of backtracking and re-guessing. For these reasons "good" Sudoku puzzles tend to have just one possible solution in each case.

From a computing perspective, the manual methods by which human players go about solving Sudoku puzzles (albeit unbeknown to most of them) closely follow simple Constraint Programming (CP) methods – each of the $n^4$ cells in the grid represents an integer variable which, initially, will have a domain of 1 through to $n^2$. Constraints can then be added in the form of "alldifferent" constraints [1] (i.e. "all of the variables in row three should have different values", etc.), and by using the pre-filled cells in the grid (e.g. "because 5 appears in row three, none of the unfilled cells in row three can contain a 5", etc.). Combinations of such constraints will reduce the domain-sizes of some of the variables and, if an appropriate propagation scheme is used, the puzzle can then (hopefully) be completed. (See the work of Simonis [2] for an example application of advanced CP techniques to logic-solvable puzzles of order-3).

It is worth noting, however, that not all puzzles will have the logic-solvable property. Indeed, Sudoku has been proved to belong to the class of NP-complete problems [3], implying that we cannot hope to find a polynomially bounded algorithm for solving *all* problem instances (unless, of course, P = NP). In other words, we can be fairly sure that there will be many problem instances where the exclusive use of logical rules will not be enough and some sort of search will also be required. For this reason, many existing automatic Sudoku solvers also include branch-and-bound search mechanisms, such as the *Sudoku Solver* by Logic.[1] However, for this sort of approach to be successful there will, of course, also be a

---

[1] Available at http://www.sudokusolver.co.uk/index.html

reliance on the search space being a manageable size. Indeed, in situations where this is not so – perhaps because the grid is still quite empty and/or because the puzzle is of a high order – then the potential timing implications of such searches might turn out to be impractical.

Given the above, in a previous paper [4] we suggested that it might also be useful to consider other types of search methods with Sudoku. Consequently, we proposed a stochastic approach based around Simulated Annealing (SA). In the next section we will describe this algorithm and its general characteristics (as reported in [4]). Subsequently, we will then suggest a way in which this algorithm might be *coupled* with a simple CP procedure to form a more powerful hybrid algorithm. In Section 3 we will then carry out a number of experiments to compare our original SA algorithm with this new approach and will discuss our results. Finally, Section 4 will conclude the paper.

## 2    A Hybrid Algorithm for Solving Sudoku

In the following descriptions, a grid cell will be described as *fixed* when its value is definitely known, either because it has been defined in the problem instance or, in the case of our hybrid algorithm, because its value has been determined by our CP procedure (to be described in Section 3). Cells whose values are undetermined will be described as *unfixed*.

The SA algorithm operates as follows. Given a problem instance of order $n$, the algorithm first creates an initial solution by assigning a value to each of the unfixed cells in the grid. This is done randomly, but in such a way so that each box ends up containing the values 1 through to $n^2$ exactly once. Creating an initial solution in this way guarantees that the third criteria of Sudoku is met; however, it also means that the grid may well feature violations of the remaining two criteria. A suitable cost function is thus:

$$\sum_{i=1}^{n^2} r(i) + \sum_{j=1}^{n^2} c(j) \tag{1}$$

where $r(i)$ and $c(j)$ represent the number of values, 1 through to $n^2$, that are *not* present in the $i$th row of cells and the $j$th column of cells respectively. An *optimal* (i.e. valid) solution will thus have a cost of zero.

In order to try and find an optimal solution, a neighbourhood operator is then used that randomly selects two unfixed cells *in the same box*, and swaps their contents. Following standard SA methods, a swap is then accepted (a) if it causes the cost to drop, or (b) with a probability $\exp(-\delta/t)$, where $\delta$ represents the proposed change in cost and $t$ is the current *temperature* of the system. During a run $t$ is slowly reduced from an initial value $t_0$ according to a geometric cooling schedule. A simple reheating function is also used that resets $t$ to $t_0$ when the algorithm considers itself to be caught in a local minimum.

Note that the neighbourhood operator ensures that the third criterion of Sudoku is always met. This means that that the total size of the search space is:

$$\prod_{i=1}^{n} \prod_{j=1}^{n} f(i,j)! \tag{2}$$

where $f(i,j)$ indicates the number of unfixed cells in the box in the $i$th row of boxes and $j$th column of boxes.

In [4], this SA algorithm was applied to a large number of solvable problem instances using a generator that was able to closely control the proportion of fixed cells in a grid (this will also be used in Section 3). Results indicated that, similarly to many other combinatorial optimisation problems (e.g. [5,6]), Sudoku also features an "easy-hard-easy" phase transition with solvable instances. In other words, the SA algorithm is generally able to discover an optimal solution when presented with instances containing very low or very high proportions of fixed cells, but at the boundary of these two extremes there occur instances that the algorithm finds more difficult to solve. The suggested reasons for this phase transition are as follows:

When the proportion of fixed cells in a grid is low, then according to eq. (2) there will be a large search space for the algorithm to navigate. However, there will also be a very large number of optimal solutions within this search space.[2] Consequently, the algorithm will nearly always be able to find one of these in a reasonable amount of time. For grids with high proportions of fixed cells, meanwhile, although there will only be a very small number of optimal solutions (and perhaps only one), the search space will be much smaller. Additionally, it is also likely that solutions to these highly constrained instances will tend to lie at the bottom of deep local minima (with a strong basin of attraction), thus also allowing easy discovery by the algorithm. However, instances at the boundary of these two extremes cause the algorithm more problems. First, the search spaces for these instances will still be relatively large, but they will also tend to admit only a small number of solutions. Second, because of their moderate numbers of constraints, the fitness landscapes will also tend to feature more plateaus and local minima, making things even more difficult for the algorithm. (See also the work of Cheeseman et al [8]).

From these explanations it is easy to see that, from the point-of-view of a stochastic search approach, an important contributing factor for an instance being "hard" is a large search space. However, it is fairly obvious that one way that we might go about alleviating this factor is by first determining the contents of as many cells as possible *before* applying such an algorithm. This is the approach that our new hybrid algorithm will take here. Given a particular problem instance, a simple CP procedure will first be applied which will fill-in and *fix* as many cells as possible. Then, once this stage has been completed the resultant partial solution will then be passed over to our original SA algorithm, which will operate in the manner that we have described.

---

[2] It has been calculated by Felgenhauer and Jarvis [7], for example, that there are 6,670,903,752,021,072,936,960 different optimal solutions for order-3 grids.

# 3   Experimental Analysis

In order to compare the performance of the two algorithms, experiments were carried out on a large number of solvable problem instances of various orders. In Section 3.1 we will describe the experiments that we conducted on randomly-formed problem instances (i.e. ones that are not necessarily *logic*-solvable). In Section 3.2 we will then present results that were gained when using collections of publicly available puzzles.

In all experiments the CP procedure that was used in conjunction with our hybrid algorithm operated by following the 5 steps given below. This procedure is deterministic.

1. For each unfixed cell in the grid, construct a list of possible values that this cell could contain by examining the contents of the cell's row, column, and box;
2. If any of these lists contains just one value, then insert this value into the cell, mark it as fixed, and go back to step 1;
3. Look at each row in turn. If any cell's list in a particular row contains a value $x$ that does not occur in any of the other cells' lists on the same row, then insert $x$ into this cell, mark the cell as fixed, and go back to step 1;
4. Repeat step 3 for each column and also each box;
5. If we are here, then the procedure cannot fix any further cells, and so end.

## 3.1   Solving Random Sudoku Grids

For our first set of experiments we used the same method of instance generation as in [4], which operates as follows:

To start with, a full and optimal Sudoku grid of a given order is taken. Such a grid can be obtained from a variety of places such as the solution pages of a Sudoku book or newspaper, by calculating the puzzles "Root Solution" (see [4]), or by simply running the SA algorithm using a blank grid as a problem instance. In the next step of the procedure, this grid is then randomly shuffled using the following five operators:

- Transpose the grid (2 possibilities);
- Permute columns of boxes within the grid ($n!$ possibilities);
- Permute rows of boxes within the grid ($n!$ possibilities);
- Permute columns of cells within a column of boxes ($n!^n$ possibilities); and
- Permute rows of cells within a column of boxes ($n!^n$ possibilities).

Note that all of these shuffle operators preserve the optimality of the grid.

Finally, a number of cells in the grid are then made blank by going through each cell in turn and deleting its contents with a probability $1 - p$, where $p$ is a parameter to be defined by the user. Obviously, this means that instances generated with a low $p$-value will have a low proportion of fixed cells (i.e. a fairly *unconstrained* problem instance), whilst larger values for $p$ will give more constrained, full problem instances.

Before comparing the SA and hybrid algorithms, it is first worth taking a look at how our CP procedure is able to cope with the randomly generated instances unaided. This is shown in fig. 2. Here, we can witness the clear pattern between the proportion of fixed cells in the initial problem instance, and the proportion of cells that are fixed after the CP procedure has been applied. As is shown, for very low $p$-values (0 to approximately 0.2) the CP procedure is not able to do anything at all, because the near-blank grids that occur here do not provide enough clues for any further cells to be filled. Meanwhile, for $p$-values of approximately 0.75 and above, because of the high proportion of fixed cells in the instances, the CP procedure is nearly always able to complete the puzzles. Finally, in-between these values, although the procedure is often unable to complete the puzzles, it is, however, usually able to fill *some* of the cells. Note that this procedure is also very quick to run – none of these trials took more than 0.03 CPU seconds (using Windows XP, with an Intel 3.2GHz processor and 1.99Gb of RAM).

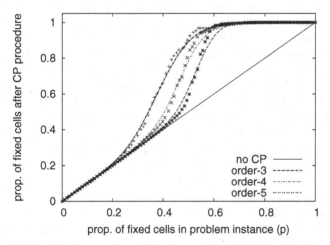

**Fig. 2.** The relationship between $p$ and the proportion of fixed cells after an application of the CP procedure for problems of order 3, 4, and 5. Each individual point in the figure is a mean, calculated after runs on twenty problem instances of a specific $p$-value. The smooth lines were produced using Gnuplot's *sbezier* function.

In order to compare the SA and hybrid algorithms directly, we used same instance generator to perform the following experiments. For $p$-values of 0 to 1.0 (incrementing in steps of 0.05), twenty separate problem instances were first created. With each of these instances, twenty separate trials with both algorithms were then performed. This was done for instances of order-3 (91 cells), order-4 (256 cells), and order-5 (625 cells), using time limits of 2, 40, and 450 CPU seconds respectively.

Finally, the SA in both algorithms was executed under the following conditions:

- An intial temperature $t_0$ was calculated by applying a small number of neighbourhood moves to the initial solution (in our case we used 100 moves). $t_0$

was then set to the variance of the cost across these moves (see [9] for the theoretical foundations of this).

- At each temperature a total of $(\sum_{i=1}^{n} \sum_{j=1}^{n} f(i,j))^2$ neighbourhood moves were attempted, where $f$ has the same interpretation as eq. 2.
- The temperature was updated using a simple geometric scheme whereby $t_{i+1} = \alpha t_i$. In our case, we set $\alpha = 0.99$.
- Finally, if no improvements in the cost were found for twenty successive temperatures, then the current temperature was reset to $t_0$, whereupon the algorithm would continue as before.

Figure 3 shows the results of these experiments and displays, for each algorithm, their *success rates* and *solution times* for all of the tested $p$-values. The success rate indicates the proportion of runs where the algorithms were able to find an optimal solution within the specified time limits. The solution time indicates the average number of CPU seconds that it took to find a solution. (In cases where the success rate was less than 1.0, those runs where optimality was not found were not considered in the latter's calculation.)

Looking at the order-3 results first, we can see that both algorithms feature a 100% success rate across all of the instances and that, in both cases, lower values for $p$ will generally require longer solution times (due to the noted fact that these instances will feature a larger search space). We can also see that for $p$-values of 0 through to 0.2, both algorithms feature roughly the same solution times. This is because, as we saw in fig. 2, with these instances the CP procedure will not usually be given sufficient clues in order to be able to fill any of the cells, and so the two algorithms are equivalent. For $p$-values of 0.25 up to around 0.7, however, we can see that the hybrid algorithm clearly shows shorter solution times, due to the fact that the CP procedure is able to fill some of the cells, therefore reducing the size of the search space for the SA algorithm.

Moving our attention onto the results of the order-4 and order-5 experiments, similar patterns also emerge with the solution times. In the centre of both graphs we also witness dips in the success rates, indicating the presence of the phase transition region that we have mentioned in Section 2. As we noted in [4], it can also be seen that as puzzle order is increased, then the effects of the phase transition also become more pronounced. Note, however, that throughout the phase transition the hybrid algorithm shows both higher success rates and also shorter solution times than the SA algorithm. According to a signed ranked test the increases in the success rates across the various $p$-values were seen to be significant (with $\geq 95\%$ confidence).

## 3.2   Solving Published Sudoku Grids

For our second set of experiments we also tested the SA and hybrid algorithms on a number of published instances that are known to have unique solutions. Our first set of order-3 puzzles was taken from the on-line resource provided by the *Los Angeles Times* [10]. These were published in the newspaper between January and March 2006 and all are known to be logic solvable. A second set of order-3

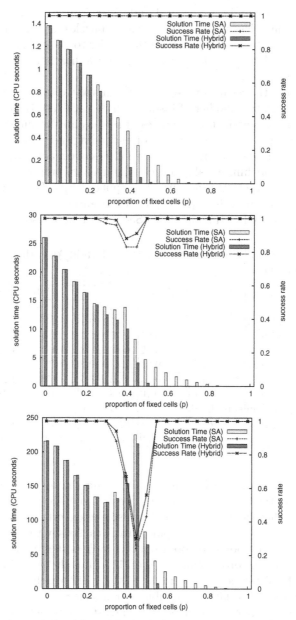

**Fig. 3.** Comparison of the SA and hybrid algorithms' performance with puzzles of order-3 (top), order-4 (middle), and order-5 (bottom)

puzzles was also taken from [11]. This resource features a very large collection of different puzzles that each contain just 17 fixed cells, which is currently the known minimum for guaranteeing that an order-3 puzzle features exactly one solution. Finally, for completeness we also used the instance generator available at [12] to produce a number of order-4 and order-5 puzzles. In general, puzzles of this size are much less popular than order-3 puzzles and so our choices here were limited. For this reason, we advise the reader to show slight caution when interpreting these latter results, as this generator has not been scientifically verified.

Table 1 contains the results of these experiments and displays the source of the puzzles, their order, the number of instances used each case (Inst.), the average proportion of fixed cells in the instances (Fixed), and the puzzles' "grades".[3] For each algorithm we then present the corresponding success rates and solution times (with standard deviation), calculated in the same manner as in Section 3.1. For the hybrid algorithm, we also present the average proportion of fixed cells that occurred *after* the CP procedure was applied (Fixed'). All entries are an average of ten runs on each of the available instances – i.e. 10 × Inst. runs in each case. The same CPU time limits as Section 3.1 were also used.

**Table 1.** Performance Comparison of the SA and Hybrid Algorithms with Instances with Unique Solutions

| Instance Description | | | | | SA Performance | | Hybrid Performance | | |
|---|---|---|---|---|---|---|---|---|---|
| Source | Order | Inst. | Fixed | Grade | Suc. Rate | Sol. Time | Fixed' | Suc. Rate | Sol. Time |
| [10] | 3 | 10 | 0.34 | gentle | 0.99 | 0.67 ± 0.1 | 0.94 | 1.00 | 0.05 ± 0.1 |
| [10] | 3 | 10 | 0.36 | tough | 0.95 | 0.76 ± 0.3 | 0.59 | 1.00 | 0.22 ± 0.2 |
| [10] | 3 | 10 | 0.34 | diabolical | 0.82 | 0.80 ± 0.3 | 0.48 | 0.99 | 0.49 ± 0.2 |
| [11] | 3 | 1000 | 0.21 | n/a | 0.01 | 1.41 ± 0.2 | 0.30 | 0.16 | 0.64 ± 0.5 |
| [12] | 4 | 10 | 0.40 | easy | 0.04 | 14.3 ± 7.5 | 0.47 | 0.24 | 25.8 ± 10.3 |
| [12] | 4 | 10 | 0.40 | hard | 0.10 | 20.3 ± 9.4 | 0.48 | 0.28 | 16.4 ± 8.6 |
| [12] | 4 | 10 | 0.49 | superhard | 0.91 | 8.64 ± 6.6 | 0.74 | 1.00 | 2.18 ± 2.5 |
| [12] | 5 | 10 | 0.46 | easy | 0.01 | 165.9 ± 0.0 | 0.51 | 0.04 | 234.1 ± 23.2 |
| [12] | 5 | 10 | 0.45 | hard | 0.00 | n/a | 0.49 | 0.00 | n/a |

As can be seen, in all cases the hybrid algorithm features an equal or higher success rate than the SA algorithm. Additionally, in all but two of the instance sets, we can see that the hybrid algorithm also gives shorter solution times (the remaining two cases are due to sampling errors caused by the very low success rates). One interesting point to note from this table is the relatively low success rates when tackling the order-3 instances of [11]. In reality these instances might be the most difficult sorts of problem for a stochastic search approach, because they feature close to the largest possible search spaces for order-3 grids whilst also ensuring that only one possible solution exists. For similar reasons, we can also see that the success rates for both algorithms also drop as the order of the puzzles is increased (with the one anomaly being the order-4 "superhard" instances, which could be due to some feature of the problem generator).

---

[3] Note that puzzle grades are probably superfluous here, because they tend to relate to the complexity of the logical techniques that a player needs to use in order to complete it. Additionally, the boundaries and adjectives that are used to define the different grades also vary from place to place.

# 4   Conclusions and Discussion

In this paper we have seen that our hybrid algorithm, which incorporates constraint programming *and* stochastic search, clearly outperforms the same stochastic search algorithm when used on its own. We have seen that the two techniques that make up this hybrid algorithm actually seem to complement one another, because it is evident that on the one hand, CP techniques have the potential to drastically improve the performance of the stochastic search algorithm, whilst on the other hand, the stochastic search algorithm can also be used to help CP-based approaches to solve a much wider range of instances (i.e. those that are not necessarily "logic-solvable"). Indeed, it is also likely that if we were to improve either aspect of the hybrid algorithm (e.g., by using more advanced deduction techniques such as the "swordfish" and "X-wing" rules [13], or by using more sophisticated search techniques), then the overall performance of the hybrid algorithm would also subsequently improve.

One interesting aspect of this work is the observation that our CP procedure allows the possibility of moving a problem instance away from the phase transition region, thus making it easier for the SA algorithm to solve. However, if this is the case, then we might ask whether it is also possible for the same procedure to move some instances *into* the phase transition region, making them *harder* to solve. We believe the answer to this question to be negative. This is because, as we have seen, the CP procedure will only ever fix a cell if it has deduced its contents with absolute certainty. Thus, although the procedure might be able to reduce the search space size by adding additional constraints, it will not reduce the number of solutions *within* this space, and its actions may well lead to a reduction in the number and/or size of plateaus in the fitness landscape. It is likely, therefore, that the instance will become easier to solve in general.

Considering future work, it is interesting to note that Sudoku can also be modelled as a graph colouring problem. This is done by considering each of the $n^4$ cells in a grid as a node, and then adding edges between any two nodes corresponding to a pair of cells in the same row, column, and/or box (meaning that the $n^2$ nodes occurring in each row/column/box will form a *clique* of size $n^2$.) Further edges can then also be added due to the pre-filled cells that are supplied with the puzzle – for example in fig. 1 it is clear that nodes (cells) 9 (top right) and 10 (first on second row) should never be the same colour, and so we can add an extra edge between these in order to ensure that they will never be allocated the same colour in a feasible solution. Given such a graph, the task is to then colour the nodes using exactly $n^2$ colours. Graph colouring has, of course, been widely studied in the past (see [14], for one example) and in the future it is likely that various techniques from this field could show applicability to Sudoku and, indeed, vice-versa.[4]

Finally, it is worth stressing that although Sudoku itself might not seem to have great practical implications in a real-world/industrial context, to its credit it

---

[4] Practitioners interested in pursuing this promising research-avenue are invited to make use of a Sudoku to graph colouring converter that we have implemented, which is available at http://www.cardiff.ac.uk/carbs/quant/rhyd/rhyd.html

is a problem that is very easy to understand, and it is certainly the case that it has encouraged many people to take an interest in constraint satisfaction problems. Perhaps more importantly though, it is noticeable that Sudoku features various similarities with other important combinatorial optimisation problems, and so its study should allow us to gain deeper insights into these as well. As an example, consider a typical timetabling problem where the aim is to assign a number of events to a limited number of timeslots and rooms in accordance with a set of constraints. In these problems it is common, among other things, to encounter *pre-assignment* constraints (e.g. "event 3 must be scheduled into room 6 in timeslot 8", etc.). In the past, various stochastic search techniques have been applied to handle these sorts of constraints in timetabling (see, for example, some of the works in [15]). However, it is noticable that this sort of constraint is actually very similar to the constraints introduced by the fixed cells in Sudoku. This suggests that it should also be useful to consider hybrid algorithms (of the sort described here) for these sorts of problems as well. Here, we refer the reader to papers by Merlot et al. [15] and Duong and Lam [16], where some preliminary work on this matter has been conducted.

# References

1. van Hoeve, W.J.: The alldifferent constraint: A survey. CoRR cs.PL/0105015 (2001)
2. Simonis, H.: Sudoku as a constraint problem. In: Hnich, B., Prosser, P., Smith, B. (eds.) Proc. 4th Int. Works. Modelling and Reformulating Constraint Satisfaction Problems, pp. 13–27 (2005)
3. Yato, T., Seta, T.: Complexity and completeness of finding another solution and its application to puzzles. IEICE Trans. Fundamentals EA6-A(5), 1052–1060 (2003)
4. Lewis, R.: In press: Metaheuristics can solve sudoku puzzles. Journal of heuristics 13 (2007)
5. Smith, B.: Phase transitions and the mushy region in constraint satisfaction problems. In: Cohn, A. (ed.) 11th European Conference on Artificial Intelligence, pp. 100–104. John Wiley and Sons ltd., Chichester (1994)
6. Turner, J.S.: Almost all $k$-colorable graphs are easy to color. Journal of Algorithms 9, 63–82 (1988)
7. Felgenhauer, B., Jarvis, F.: Mathematics of sudoku. Online Resource: (2006), http://www.afjarvis.staff.shef.ac.uk/sudoku/
8. Cheeseman, P., Kanefsky, B., Taylor, W.M.: Where the really hard problems are. In: IJCAI-91. Proceedings of the Twelfth International Joint Conference on Artificial Intelligence, Sidney, Australia, pp. 331–337 (1991)
9. van Laarhoven, P., Aarts, E.: Simulated Annealing: Theory and Applications. D. Reidel Publishing Company, Dordrecht (1987)
10. Mepham, M.: Sudoku archive. Online (2006), http://www.sudoku.org.uk/backpuzzles.htm
11. Royle, G.: Minimum sudoku. Online (2006), http://www.csse.uwa.edu.au/~gordon/sudokumin.php
12. Hanssen, V.: Sudoku puzzles. Online (2006), http://www.menneske.no/sudoku/eng/

13. Armstrong, S.: Sudoku solving techniques. Online (2006),
    http://www.sadmansoftware.com/sudoku/techniques.htm
14. Jensen, T.R., Toft, B.: Graph Coloring Problems, 1st edn. Wiley-Interscience,
    Chichester (1994)
15. Burke, E.K., De Causmaecker, P. (eds.): PATAT 2002. LNCS, vol. 2740. Springer,
    Heidelberg (2003)
16. Duong, T.A., Lam, K.H.: Combining constraint programming and simulated an-
    nealing on university exam timetabling. In: RIVF2004. Proceedings of the 2nd
    International Conference in Computer Sciences, Research, Innovation & Vision for
    the Future, pp. 205–210 (2004)

# Improvement Strategies for
# the F-Race Algorithm:
# Sampling Design and Iterative Refinement

Prasanna Balaprakash, Mauro Birattari, and Thomas Stützle

IRIDIA, CoDE, Université Libre de Bruxelles, Brussels, Belgium
{pbalapra,mbiro,stuetzle}@ulb.ac.be

**Abstract.** Finding appropriate values for the parameters of an algorithm is a challenging, important, and time consuming task. While typically parameters are tuned by hand, recent studies have shown that
automatic tuning procedures can effectively handle this task and often
find better parameter settings. F-Race has been proposed specifically for
this purpose and it has proven to be very effective in a number of cases.
F-Race is a racing algorithm that starts by considering a number of candidate parameter settings and eliminates inferior ones as soon as enough
statistical evidence arises against them. In this paper, we propose two
modifications to the usual way of applying F-Race that on the one hand,
make it suitable for tuning tasks with a very large number of initial
candidate parameter settings and, on the other hand, allow a significant
reduction of the number of function evaluations without any major loss
in solution quality. We evaluate the proposed modifications on a number
of stochastic local search algorithms and we show their effectiveness.

## 1   Introduction

The full potential of a parameterized algorithm cannot be achieved unless its
parameters are fine tuned. Often, practitioners tune the parameters using their
personal experience guided by some rules of thumb. Usually, such a procedure is
tedious and time consuming and, hence, it is not surprising that some authors
say that 90% of the total time needed for developing an algorithm is dedicated
to find the right parameter values [1]. Therefore, an effective automatic tuning
procedure is an absolute must by which the computational time and the human
intervention required for tuning can be significantly reduced. In fact, the selection of parameter values that drive heuristics is itself a scientific endeavor and
deserves more attention than it has received in the operations research literature [2]. In this context, few procedures have been proposed in the literature.
F-Race [3,4] is one among them and has proven to be successful and useful in a
number of tuning tasks [4,5,6,7].

Inspired by a class of racing algorithms proposed in the machine learning literature, F-Race evaluates a given set of parameter configurations sequentially on
a number of problem instances. As soon as statistical evidence is obtained that

T. Bartz-Beielstein et al. (Eds.): HM 2007, LNCS 4771, pp. 108–122, 2007.

a candidate configuration is worse than at least another one, the inferior candidate is discarded and not considered for further evaluation. In all previously published works using F-Race, the initial candidate configurations were obtained through a full factorial design. This design is primarily used to select the best parameter configuration from a relatively small set of promising configurations that the practitioner has already established. Nevertheless, the main difficulty of this design is that, if the practitioner is confronted with a large number of parameters and a wide range of possible values for each parameter, the number of initial configurations becomes quite large. In such cases, the adoption of the full factorial design within F-Race can become impractical and computationally prohibitive. In order to tackle this problem, we propose two modifications to the original F-Race approach. The first consists in generating configurations by random sampling. Notwithstanding the simplicity, the empirical results show that this approach can be more effective—in the context of tuning tasks—than the adoption of a full factorial design. However, if the number of parameters is large, this methodology might need a large number of configurations to achieve good results. We alleviate this problem taking inspiration from model-based search techniques [8]. The second procedure uses a probabilistic model defined on the set of all possible parameter configurations and at each iteration, a small set of parameter configurations is generated according to the model. Elite configurations selected by F-Race are then used to update the model in order to bias the search around the high quality parameter configurations.

The paper is organized as follows: In Section 2, we introduce the proposed approach and we present some empirical results in Section 3. We discuss some related work in Section 4, and conclude the paper in Section 5.

## 2   Sampling F-Race and Iterative F-Race for Tuning Stochastic Local Search Algorithms

For a formal definition of the problem of tuning SLS algorithms, we follow Birattari et al. [3]. The problem is defined as a 6 tuple $\langle \Theta, I, P_I, P_c, t, \mathcal{C} \rangle$, where $\Theta$ is the finite set of candidate configurations, $I$ is the possibly infinite set of problem instances, $P_I$ is a probability measure over the set $I$, $t$ is a function associating to every instance the computation time that is allocated to it, $P_C$ is a probability measure over the set of all possible values for the cost of the best solution found in a run of a configuration $\theta \in \Theta$ on an instance $i$, $\mathcal{C}(\theta)$ is the criterion that needs to be optimized with respect to $\theta$: the solution of the tuning problem consists in finding a configuration $\theta^*$ such that

$$\theta^* = \arg \min_{\theta} \mathcal{C}(\theta). \tag{1}$$

Typically, $\mathcal{C}(\theta)$ is an expected value where the expectation is considered with respect to both $P_I$ and $P_C$. The main advantage of using expectation is that it can be effectively and reliably estimated with Monte Carlo procedures. In this

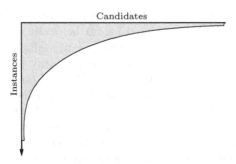

**Fig. 1.** Visual representation of a typical trace of F-Race giving the number of surviving configurations in dependence of the number of instances seen. The *x-axis* represents the number of candidate configurations that are still in the race and the *y-axis* represents the number of instances that has been used for evaluating the configurations. As the evaluation proceeds, F-Race focuses more and more on the promising configurations.

paper, we focus on the minimization of the expected value of the solution cost and the criterion is given as:

$$\mathcal{C}(\theta) = \mathbb{E}_{I,C}\Big[c(\theta, i)\Big] = \int_I \int_C c_t(\theta, i) \, \mathrm{d}P_C(c_t|\theta, i) \, \mathrm{d}P_I(i), \qquad (2)$$

where, $c_t(\theta, i)$ is a random variable that represents the cost of the best solution found by running configuration $\theta$ on instance $i$ for $t$ seconds. The integration is taken in the Lebesgue sense and the integrals are estimated in a Monte Carlo fashion on the basis of a so-called *tuning set* of instances. It is straightforward to use criteria other than the expected value such as inter-quartile range of the solution cost. In the case of decision problems, the practitioner might be interested in minimizing the run-time of an algorithm, a task that can be handled in a straightforward way by F-Race.

F-Race is inspired by a class of racing algorithms proposed in the machine learning literature for tackling the model selection problem [9,10]. In F-Race, as in other racing algorithms, a set of given candidate configurations are sequentially evaluated on a number of tuning instances. As soon as sufficient evidence is gathered that a candidate configuration is worse than at least another one, the former is discarded from the race and is not further evaluated. The race terminates when either one single candidate configuration remains, or the available budget of computation time is used. The peculiarity of F-Race compared to other racing algorithms is the adoption of the *Friedman two-way analysis of variance by ranks* [11], a nonparametric statistical test that appears particularly suitable in the context of racing algorithms for the tuning task. The progress of the F-Race procedure can be graphically illustrated as shown in Figure 1.

The main focus of this paper is the method by which the initial set of configurations is obtained in F-Race: while F-Race does not specify how $\Theta$ is defined, in most of the studies on F-Race, the configurations are defined using a full factorial design (FFD). In the simplest case, this is done as follows: Let M = $\{M_1, \ldots, M_d\}$ be the set of parameters that need to be tuned whose ranges are given by $(min_k, max_k)$, for $k = 1, \ldots, d$, where $min_k$ and $max_k$ are the minimum and maximum values of the parameter $M_k$, respectively. For each element in M, the practitioner has to choose a certain number of values; each possible combination of these parameter values leads to one unique configuration and the set of all possible combinations forms the initial set of configurations. If $l_k$ values are chosen for $M_k$, then the number of initial configurations is $\prod_{k=1}^{d} l_k$. When each parameter takes $l$ values, then $\prod_{k=1}^{d} l = l^d$; that is, the number of configurations grows exponentially with respect to the number of parameters. As a consequence, even a reasonable number of possible values for each parameter makes the adoption of a full factorial design impractical and computationally prohibitive.

## 2.1  Sampling F-Race

A simple way to overcome the shortcomings of FFD is sampling. This means that the elements of $\Theta$ are sampled according to a given probability measure $P_X$ defined on the space $X$ of parameter values. If *a priori* knowledge is available on the effect of the parameters and on their interactions, this knowledge can be used to shape the probability measure $P_X$ and therefore to suitably bias the sampling of the initial configurations. On the other hand, if no *a priori* knowledge on the parameter values is available, except the boundary constraints, then each possible value in the available range for each parameter should be given equal probability of being selected in sampling. In this case, $P_X$ is a $d$-variate uniform distribution, which is factorized by a product of $d$ univariate independent uniform distributions. A sample from the $d$-variate uniform distribution is a vector corresponding to a configuration $\theta$ such that a value $x_k$ in the vector is sampled from the univariate independent uniform distribution parameterized by $(min_k, max_k)$. We call this strategy *random sampling design* (RSD). The F-Race procedure is then applied to the set of sampled configurations. We denote this procedure as RSD/F-Race. It should be noted that the performance of the winning configuration is greatly determined by the number of sampled configurations, $N_{max}$.

## 2.2  Iterative F-Race

RSD/F-Race can identify promising configurations in the search space. However, finding the best configuration from the promising regions is often a difficult task. In order to address this issue, we propose iterative F-Race (I/F-Race), a supplementary mechanism to the original F-Race approach. It is an iterative procedure in which each iteration consists in first defining a probability measure over the parameter space using promising configurations obtained from the previous iteration, then generating configurations that are distributed according to the newly defined probability measure, and finally applying F-Race on the generated configurations. This approach falls under the general framework of model-based search [8].

The way in which the probability measure is defined at each iteration plays a crucial role in biasing the search towards regions containing high quality configurations. The main issues in the search bias are the choice of the distribution and search intensification. For what concerns the distribution, there exist a number of choices. Here, we adopt a $d$-variate normal distribution parameterized by mean vector and covariance matrix. In order to intensify the search around the promising configurations, a $d$-variate normal distribution is defined on each surviving configuration from the previous iteration such that the distribution is centered at the values of the corresponding configuration. Moreover, the spread of the normal densities given by the covariance matrix is gradually reduced at each iteration.

This paper focuses on a scenario in which the practitioner does not have any *a priori* knowledge on the parameter values. Hence, we assume that the values taken by the parameters are independent, that is, knowing a value for a particular parameter does not give any information on the values taken by the other parameters. Consequently, the $d$-variate normal distribution is factorized by a product of $d$ univariate independent normal densities parameterized by $\mu = (\mu_1, \ldots, \mu_d)$ and $\sigma = (\sigma_1, \ldots, \sigma_d)$. At each iteration, the standard deviation vector $\sigma$ of the normal densities is reduced heuristically using the idea of volume reduction: Suppose that $N_s$ configurations survive after a given iteration; we denote the surviving configurations as $\theta_s = (x_1^s, \ldots, x_d^s)$, for $s = 1, \ldots, N_s$. At a given iteration $r$, let $V_r$ be the total volume of the $d$-dimensional sampling region bounded by $(\mu_k^{s_r} \pm \sigma_k^{s_r})$, for $k = 1, \ldots, d$; for iteration $r + 1$, in order to intensify the search, we reduce the volume of the sampling region by a factor equal to the number of sample configurations allowed for each iteration, $N_{max}$; therefore $V_{r+1} = V_r / N_{max}$, from which after some basic mathematical transformation, we have:

$$\sigma_k^s = R_k^{s_{prev}} \cdot \left( \frac{1}{N_{max}} \right)^{1/d} \qquad \text{for } k = 1, \ldots, d, \qquad (3)$$

where $R_k^{s_{prev}}$ is set to standard deviation of the normal distribution component from which $x_k^s$ has been sampled from the previous iteration. In simple terms, the adoption of Equation 3 allows I/F-Race to reduce the range of each parameter that falls around one standard deviation from the mean at a constant rate of $(1/N_{max})^{1/d}$ for each iteration—the larger the value of $N_{max}$, the higher the rate of volume reduction. Though one could use more advanced techniques to update the distribution as suggested by the model-based search framework [8], we have adopted the above described heuristic way of intensifying search due to its simplicity.

Note that in the first iteration, a $d$-variate uniform distribution is used as the probability measure, thus for the following iteration, $R_k^{s_{prev}}$ is set to the half of range, that is, $(max_k - min_k)/2$, where $max_k$ and $min_k$ are parameters of the uniform distribution component from which $x_k^s$ has been sampled, respectively.

The proposed approach adopts a strategy in which the number of configurations drawn from a $d$-variate normal distribution defined on a surviving configuration is inversely proportional to the configurations' expected solution cost. Recall that we are faced with the minimization of the expected solution cost. To do so, a *selection probability* is defined: the surviving configurations are ranked according to their expected solution costs and the probability of selecting a $d$-variate normal distribution defined on a configuration with rank $z$ is given by

$$p_z = \frac{N_s - z + 1}{N_s \cdot (N_s + 1)/2}. \tag{4}$$

A configuration is obtained by first choosing a $d$-variate normal distribution according to Equation 4, and then sampling from the chosen distribution. This is repeated until $N_{max}$ configurations are sampled.

**Implementation Specific Details.** In order to guarantee that I/F-Race does a specific minimum number of iterations and that it has a minimum number of survivors, we have modified F-Race slightly to stop it prematurely. At each iteration, the race is stopped if one of the following conditions is true:

- when $N_{min}$ configurations remain;
- when a certain amount of computational budget, $CB_{min}$, is used;
- when the configurations in the race are evaluated on at least $I_{max}$ instances.

Though these modifications introduce 3 parameters, they are set in a reasonable and straightforward way with respect to the total computational budget $CB$ when the algorithm starts: (i) $CB_{min}$ is set to $CB/5$: this setting allows I/F-Race to perform at least five iterations; (ii) $N_{min}$ is set to $d$: this setting enables I/F-Race to search in a number of promising regions rather than just concentrating on a single region; (iii) $I_{max}$ is set to $2 \cdot (CB_{min}/N_{max})$: if none of the configurations is eliminated from the race then each configuration has been evaluated on $CB_{min}/N_{max}$ instances; hence, twice this value seems to be a reasonable upper bound.

The maximum number $N_{max}$ of configurations allowed for each race is kept constant throughout the procedure. Moreover, the $N_s$ configurations that have survived the race are allowed to compete with the newly sampled configurations. Therefore, $N_{max} - N_s$ configurations are sampled anew at each iteration.

The order in which the instances are given to the race is randomly shuffled for each iteration. Since the surviving configurations of each race are allowed to enter into the next race, their results could be reused if the configuration has already been evaluated on a particular instance. However, since we do not want to bias I/F-Race in the empirical study, we did not use this possibility here.

The boundary constraints are handled in an explicit way. We adopt a method that consists in assigning the boundary value if the sampled value is outside the boundary. The rationale behind this adoption is to allow the exploration of values that lay at the boundary. In the case of parameters that take integer values, the value assigned to each integer parameter in the entire procedure is rounded off to the nearest integer.

## 3    Experiments

In this section, we study the proposed RSD/F-Race and I/F-Race using three examples. Though any parameterized algorithm may be tuned, all three examples concern the tuning of stochastic local search algorithms [12]: (i) tuning $\mathcal{MAX} - \mathcal{MIN}$ ant system ($\mathcal{MMAS}$) [13], a particular ant colony optimization algorithm, for a class of instances of the TRAVELING SALESMAN PROBLEM (TSP), (ii) tuning estimation-based local search, a new local search algorithm for stochastic combinatorial optimization problems [14], for a class of instances of the PROBABILISTIC TRAVELING SALESMAN PROBLEM (PTSP), and (iii) tuning a simulated annealing algorithm for a class of instances of the VEHICLE ROUTING PROBLEM WITH STOCHASTIC DEMANDS (VRP-SD). The primary goal of these examples is to show that RSD/F-Race and I/F-Race can significantly reduce the computational budget required for tuning.

We compare RSD/F-Race and I/F-Race with an implementation of F-Race that uses a full factorial design (FFD). For RSD/F-Race and I/F-Race we make the assumption that the *a priori* knowledge on the parameter values is not available. In the case of FFD, we consider two variants:

1. FFD that uses *a priori* knowledge; a parameter $M_k$ is allowed to take $l_k$ values, for $k = 1, \ldots, d$, where $l_k$ values are chosen according to the *a priori* knowledge available on the parameter values; we denote this variant by FFD$_A$/F-Race.
2. FFD that uses random values: a parameter $M_k$ is allowed to take $l_k$ values, for $k = 1, \ldots, d$, where $l_k$ values are chosen randomly; we denote this variant by FFD$_R$/F-Race. Note that the number of configurations in this variant is the same as that of FFD$_A$/F-Race. This serves as a yardstick to analyze the usefulness of the *a priori* knowledge. The rationale behind the adoption of this yardstick is that if one just takes random values for FFD and achieves better results then FFD$_A$/F-Race, then we can conjecture that the available *a priori* knowledge is either not accurate or simply not useful, at least in the examples that we consider here.

The minimum number of steps allowed in F-Race for all algorithms before applying the *Friedman* test is set to 5 as proposed in [4].

The maximum computational budget of FFD$_A$/F-Race and FFD$_R$/F-Race are set to 10 times the number of initial configurations. The rationale behind this choice is that, if none of the configurations is eliminated, FFD$_A$/F-Race and FFD$_R$/F-Race evaluate all the configurations on at least 10 instances. This budget is also given for RSD/F-Race and I/F-Race. In order to force RSD/F-Race to use the entire computational budget, the number of configurations is set to one-sixth of the computational budget. Since I/F-Race needs to perform at least five F-races with the same budget as that of RSD/F-Race, the number of initial configurations in each F-Race run by I/F-Race is set to one-fifth of the number of configurations given to RSD/F-Race. Moreover, in order to study the effectiveness of RSD/F-Race and I/F-Race under strong budget constraints, the computational budget is reduced by a factor of two, four, and eight. Note that, in these cases, the number of configurations in RSD/F-Race and I/F-Race is set according to the allowed budget using the same rule as described before.

Each tuning algorithm is allowed to perform 10 trials and the order in which the instances are given to an algorithm is randomly shuffled for each trial.

All tuning algorithms were implemented and run under R version $2.4^1$ and we used a public domain implementation of F-Race in R which is freely available for download [15]. $\mathcal{MMAS}^2$ and estimation-based local search were implemented in C and compiled with gcc, version 3.4. Simulated annealing for VRP-SD is implemented in C++. Experiments were carried out on AMD Opteron$^{TM}$244 1.75 GHz processors with 1 MB L2-Cache and 2 GB RAM, running under the Rocks Cluster Distribution 4.2 GNU/Linux.

In order to quantify the effectiveness of each algorithm, we study the expected solution cost of the winning configuration $\mathcal{C}(\theta^*)$, where the expectation is taken with respect to the set of all trials and the set of all test instances. We report the expected solution cost of each algorithm, measured as the percentage deviation from a *reference cost*, which is given by the average over $\mathcal{C}(\theta^*)$ obtained by each algorithm. The adoption of *reference cost* allows us to compare the expected solution cost of different algorithms more directly.

In order to test whether the observed differences between the expected solution costs of different tuning algorithms are significant in a statistical sense, a random permutation test is adopted. The level of significance at which we reject the null hypothesis is 0.05; two sided $p$-value is computed for each comparison.

### 3.1   Tuning $\mathcal{MMAS}$ for TSP

In this study, we tune 6 parameters of $\mathcal{MMAS}$:

1. relative influence of pheromone trails, $\alpha$;
2. relative influence of heuristic information, $\beta$;
3. pheromone evaporation rate, $\rho$;
4. parameter used in computing the minimum pheromone trail value $\tau_{min}$, $\gamma$, which is given by $\tau_{max}/(\gamma * instance\_size)$;
5. number of ants, $m$;
6. number of neighbors used in the solution construction phase, $nn$.

In FFD$_A$/F-Race and FFD$_R$/F-Race, each parameter is allowed to take 3 values. The parameter values in FFD$_A$/F-Race are set as follows: $\alpha \in \{0.75, 1.00, 1.50\}$, $\beta \in \{1.00, 3.00, 5.00\}$, $\rho \in \{0.01, 0.02, 0.03\}$, $\gamma \in \{1.00, 2.00, 3.00\}$, $m \in \{500, 750, 1000\}$, and $nn \in \{20, 30, 40\}$. These values are chosen reasonably close to the values proposed in [16]. Note that the values are chosen from the version without the local search. Table 1 shows the ranges of the parameters considered for RSD/F-Race and I/F-Race. The computational time allowed for evaluating a configuration on an instance is set to 20 seconds. Instances are generated with the DIMACS instance generator [17]. We used uniformly distributed Euclidean instances of size 750; 1000 instances were generated for tuning; 300 other instances were generated for evaluating the winning configuration. Table 2 shows

---

[1] R is a language and environment for statistical computing that is freely available under the GNU GPL license at http://www.r-project.org/

[2] We used the ACOTSP package, which is a public domain software that provides an implementation of various ant colony optimization algorithms applied to the symmetric TSP. The package available at: http://www.aco-metaheuristic.org/aco-code/

**Table 1.** Ranges of the parameter values considered for tuning $\mathcal{MMAS}$ for TSP with RSD/F-Race and I/F-Race

| parameter | range |
|:---:|:---:|
| $\alpha$ | $[0.0, 1.5]$ |
| $\beta$ | $[0.0, 5.0]$ |
| $\rho$ | $[0.0, 1.0]$ |
| $\gamma$ | $[0.01, 5.00]$ |
| $m$ | $[1, 1200]$ |
| $nn$ | $[5, 50]$ |

**Table 2.** Computational results for tuning $\mathcal{MMAS}$ for TSP. The column entries with the label per.dev shows the percentage deviation of each algorithms' expected solution cost from the *reference cost*: $+x$ means that the expected solution cost of the algorithm is $x\%$ more than the *reference cost* and $-x$ means that the expected solution cost of the algorithm is $x\%$ less than the *reference cost*. The column entries with the label with max.bud shows the maximum number of evaluations given to each algorithm and the column with the label usd.bud shows the average number of evaluations used by each algorithm.

| algo | per.dev | max.bud | usd.bud |
|:---|:---:|:---:|:---:|
| FFD$_R$/F-Race | $+13.45$ | 7290 | 5954 |
| FFD$_A$/F-Race | $+11.13$ | 7290 | 5233 |
| RSD/F-Race | $-2.69$ | 7290 | 7232 |
| I/F-Race | $-3.92$ | 7290 | 7181 |
| RSD/F-Race | $-2.55$ | 3645 | 3275 |
| I/F-Race | $-3.84$ | 3645 | 3564 |
| RSD/F-Race | $-2.51$ | 1822 | 1699 |
| I/F-Race | $-3.66$ | 1822 | 1793 |
| RSD/F-Race | $-2.17$ | 911 | 823 |
| I/F-Race | $-3.23$ | 911 | 894 |

the percentage deviation of each algorithms' expected solution cost from the *reference cost*, the maximum budget allowed for each algorithm and the average number of evaluations used by each algorithm.

From the results, we can see that I/F-Race is very competitive: under equal computational budget, the expected solution cost of I/F-Race is approximately 17% and 15% less than that of FFD$_R$/F-Race and FFD$_A$/F-Race, respectively (the observed differences are significant according to the random permutation test). On the other hand, the expected solution cost of RSD/F-Race is also very low. However, I/F-Race reaches an expected cost that is about 1% less than that of RSD/F-Race. Indeed, the observed difference is significant in a statistical sense. Regarding the budget, FFD$_R$/F-Race and FFD$_A$/F-Race use only 80% and 70% of the maximum budget. This early termination of the F-Race is attributed to the adoption of FFD: since, there are rather few possible values for each parameter, the inferior configurations are identified and discarded within few steps.

**Table 3.** Ranges of the parameter values considered for tuning estimation-based local search for PTSP with RSD/F-Race and I/F-Race

| parameter | range |
|-----------|-------|
| $p_1$ | $[0.0, 1.0]$ |
| $w$ | $[0, 100]$ |
| $p_2$ | $[0.0, 1.0]$ |

However, the poor performance of FFD$_R$/F-Race and FFD$_A$/F-Race is not only attributable to the fact that they do not use the budget effectively: Given only half of the computational budget (a maximum budget of 3645), RSD/F-Race and I/F-Race achieve expected solution costs that are still 17% and 15% lower than FFD$_R$/F-Race and FFD$_A$/F-Race, respectively (the observed differences are significant according to the random permutation test). Another important observation is that, in the case of I/F-Race and RSD/F-Race, reducing the budget does not degrade the effectiveness to a large extent. Furthermore, in all these reduced budget cases, I/F-Race achieves an expected solution cost which is approximately 1% less than that of RSD/F-Race (the observed differences are significant according to the random permutation test).

## 3.2 Tuning Estimation-Based Local Search for PTSP

Estimation-based local search is an iterative improvement algorithm that makes use of the 2-exchange and node-insertion neighborhood relation, where the delta evaluation is performed using empirical estimation techniques [14]. In order to increase the effectiveness of this algorithm, a variance reduction technique called importance sampling has been adopted. Three parameters that need to be tuned in this algorithm are:

1. shift probability for 2-exchange moves, $p_1$;
2. number of nodes allowed for shift in 2-exchange moves, $w$;
3. shift probability for node-insertion moves, $p_2$.

Since this is a recently developed algorithm, *a priori* knowledge is not available on the parameter values. Thus, in FFD$_A$/F-Race, the values are assigned by discretization: for each parameter, the range is discretized as follows: $p_1 = p_2 \in \{0.16, 0.33, 0.50, 0.66, 0.83\}$, and $w = \{8, 17, 25, 33, 42\}$. Table 3 shows the ranges of the parameters considered for RSD/F-Race and I/F-Race. Estimation-based local search is allowed to run until it reaches a local optimum. Instances are generated as described in [14]: we used clustered Euclidean instances of size 1000; 800 instances were generated for tuning; 800 more instances were generated for evaluating the winning configuration.

The computational results show that the difference between the expected cost of the solutions obtained by different algorithms exhibits a trend similar to the one observed in the TSP experiments. However, the percentage deviations from

**Table 4.** Computational results for tuning estimation-based local search for PTSP. The column entries with the label `per.dev` shows the percentage deviation of each algorithms' expected solution cost from the *reference cost*: $+x$ means that the expected solution cost of the algorithm is $x\%$ more than the *reference cost* and $-x$ means that the expected solution cost of the algorithm is $x\%$ less than the *reference cost*. The column entries with the label with `max.bud` shows the maximum number of evaluations given to each algorithm and the column with the label `usd.bud` shows the average number of evaluations used by each algorithm.

| algo | per.dev | max.bud | usd.bud |
|------|---------|---------|---------|
| FFD$_R$/F-Race | $+1.45$ | 1250 | 1196 |
| FFD$_A$/F-Race | $+1.52$ | 1250 | 1247 |
| RSD/F-Race | $-0.62$ | 1250 | 1140 |
| I/F-Race | $-0.53$ | 1250 | 1232 |
| RSD/F-Race | $-0.17$ | 625 | 615 |
| I/F-Race | $-0.52$ | 625 | 618 |
| RSD/F-Race | $-0.06$ | 312 | 307 |
| I/F-Race | $-0.58$ | 312 | 278 |
| RSD/F-Race | $-0.37$ | 156 | 154 |
| I/F-Race | $-0.11$ | 156 | 150 |

the *reference cost* are relatively small: under equal computational budget, the expected solution cost of I/F-Race and RSD/F-Race are approximately 2% less than that of FFD$_R$/F-Race and FFD$_A$/F-Race, respectively. Note that this difference is significant according to a random permutation test. Though RSD/F-Race obtains an expected solution cost which is 0.01% less than that of I/F-Race, the random permutation test does not reject the null hypothesis. The overall low percentage deviation between algorithms is attributed to the fact that the estimation based local search is not extremely sensitive to the parameter values: there are only 3 parameters and interactions among them are quite low. As a consequence, the tuning task becomes relatively easy (as in the case of the previous task of tuning of $\mathcal{MMAS}$). This can be easily seen with the used budget of FFD$_R$/F-Race: if the task of finding good configurations were difficult, the race would have terminated early. Yet, this is not the case and almost the entire computational budget has been used.

The numerical results on the budget constraints show that both RSD/F-Race and I/F-Race are indeed effective. Given only one-eighth of the computational budget (a maximum budget of 156 evaluations), RSD/F-Race and I/F-Race achieve expected solution costs which are approximately 1.4% less than that of FFD$_R$/F-Race and FFD$_A$/F-Race. This observed difference is significant according to the random permutation test. However, in this case, the random permutation test cannot reject the null hypothesis that RSD/F-Race and I/F-Race achieve expected solution costs that are equivalent. On the other hand, given one-half and one-fourth of the computational budget, I/F-Race achieves an expected solution cost that is approximately 0.4% less that of RSD/F-Race (observed differences are significant according to the random permutation test).

**Table 5.** Ranges of the parameter values considered for tuning a simulated annealing algorithm for VRP-SD with RSD/F-Race and I/F-Race

| parameter | range |
|:---:|:---:|
| $\alpha$ | $[0.0, 1.0]$ |
| $q$ | $[1, 100]$ |
| $r$ | $[1, 100]$ |
| $f$ | $[0.01, 1.00]$ |

**Table 6.** Computational results for tuning a simulated annealing algorithm for VRP-SD. The column entries with the label per.dev shows the percentage deviation of each algorithms' expected solution cost from the *reference cost*: $+x$ means that the expected solution cost of the algorithm is $x\%$ more than the *reference cost* and $-x$ means that the expected solution cost of the algorithm is $x\%$ less than the *reference cost*. The column entries with the label with max.bud shows the maximum number of evaluations given to each algorithm and the column with the label usd.bud shows the average number of evaluations used by each algorithm.

| algo | per.dev | max.bud | usd.bud |
|:---|:---:|:---:|:---:|
| FFD$_R$/F-Race | $+0.02$ | 810 | 775 |
| FFD$_A$/F-Race | $+0.11$ | 810 | 807 |
| RSD/F-Race | $-0.05$ | 810 | 804 |
| I/F-Race | $-0.03$ | 810 | 797 |
| RSD/F-Race | $-0.03$ | 405 | 399 |
| I/F-Race | $-0.05$ | 405 | 399 |
| RSD/F-Race | $+0.02$ | 202 | 200 |
| I/F-Race | $-0.01$ | 202 | 200 |
| RSD/F-Race | $+0.02$ | 101 | 101 |
| I/F-Race | $+0.02$ | 101 | 100 |

### 3.3   Tuning a Simulated Annealing Algorithm for VRP-SD

In this study, 4 parameters of a simulated annealing algorithm have been tuned:

1. cooling rate, $\alpha$;
2. a parameter used to compute the number of iterations after which the process of reheating can be applied, $q$;
3. another parameter used to compute the number of iterations after which the process of reheating can be applied, $r$;
4. parameter used in computing the starting temperature value, $f$;

In FFD$_A$/F-Race and FFD$_R$/F-Race, each parameter is allowed to take 3 values and in the former, the values are chosen close to the values adopted in [7]: $\alpha \in \{0.25, 0.50, 0.75\}$, $q \in \{1, 5, 10\}$, $r \in \{20, 30, 40\}$, $f \in \{0.01, 0.03, 0.05\}$. Table 5 shows the ranges of the parameters considered for RSD/F-Race and I/F-Race. In all algorithms, the computational time allowed for evaluating a configuration on an instance is set to 10 seconds. Instances are generated as described in [7]; 400 instances were generated for tuning; 200 more instances were generated for evaluating the winning configuration.

The computational results show that, similar to the previous example, the tuning task is rather easy. Concerning the expected solution cost, the randomized permutation test cannot reject the null hypothesis that the different algorithms produce equivalent results. However, it should be noted that the main advantage of RSD/F-Race and I/F-Race is their effectiveness under strong budget constraints: RSD/F-Race and I/F-Race, given only one-eighth of the computational budget, achieve expected solution costs that are not significantly different from FFD$_R$/F-Race and FFD$_A$/F-Race.

## 4   Related Work

The problem of tuning SLS algorithm is essentially a mixed variable stochastic optimization problem. Even though a number of algorithms exist for mixed variable stochastic optimization, it is quite difficult to adopt them for tuning. The primary obstacle is that, since these algorithms have parameters, tuning them is indeed paradoxical. Few procedures have been developed specifically for tuning algorithms: Kohavi and John [18] proposed an algorithm that makes use of best-first search and cross-validation for automatic parameter selection. Boyan and Moore [19] introduced a tuning algorithm based on machine learning techniques. The main emphasis of these two works is given only to the parameter value selection; there is no empirical analysis of these algorithms when applied to large number of parameters that have wide range of possible values. Audet and Orban [20] proposed a pattern search technique called mesh adaptive direct search that uses surrogate models for algorithmic tuning. In this approach, a conceptual mesh is constructed around a solution and the search for better solutions is done around this mesh. The surrogates are used to reduce the computation time by providing an approximation to the original response surface. Nevertheless, this approach has certain number of parameters and it has never been used for tuning SLS algorithms. Adenso-Diaz and Laguna [1] designed an algorithm called CALIBRA specifically for fine tuning SLS algorithms. It uses Taguchi's fractional factorial experimental designs coupled with local search. In this work, the authors explicitly mention that tuning a wide range of possible values for parameters is feasible with their algorithm. However, a major limitation of this algorithm is that one cannot use it for tuning SLS algorithms with more than five parameters. Beielstein et al. [21] proposed an approach to reduce the difficulty of the tuning task. This approach consists in first identifying the parameters that have a significant impact on the algorithms' performance through sensitivity analysis and then tuning them. Recently, Hutter et al. [22] proposed an iterated local search algorithm for parameter tuning called paramILS. This algorithm is shown to be very effective and most importantly, it can be used to tune algorithms with a large number of parameters.

# 5    Conclusions and Future Work

We proposed two supplementary procedures for F-Race that are based on random sampling, RSD/F-Race, and model-based search techniques, I/F-Race. While the adoption of full factorial design in the F-Race framework is impractical and computationally prohibitive when used to identify the best from a large number of parameter configurations, RSD/F-Race and I/F-Race are useful in such cases. Since the proposed approaches are quite effective under strong budget constraints, they can reduce significantly the computational time required for tuning. However, based on the case studies, we conjecture that the expected solution cost obtained by RSD/F-Race and I/F-Race is mainly attributed to the difficulty of the tuning task.

Concerning the future research, we will extend our approach to include categorical variables. Regarding I/F-Race, we will also investigate the adoption of distributions like Cauchy and some advanced techniques for updating the distribution. Finally, from the case studies that were made in the paper, we speculate that the difficulty of the tuning task depends on a number of factors such as the sensitivity of the parameters that need to be tuned and problem instances that need to be tackled. In this context, search space analysis on the parameter values is an area to investigate further.

**Acknowledgments.** This research has been supported by COMP$^2$SYS, a Marie Curie Early Stage Research Training Site funded by the European Community's Sixth Framework Programme under contract number MEST-CT-2004-505079, and by the ANTS project, an *Action de Recherche Concertée* funded by the Scientific Research Directorate of the French Community of Belgium. Prasanna Balaprakash and Thomas Stützle acknowledge support from the Belgian FNRS of which they are an Aspirant and a Research Associate, respectively. The information provided is the sole responsibility of the authors and does not reflect the opinion of the sponsors. The European Community is not responsible for any use that might be made of data appearing in this publication.

# References

1. Adenso-Diaz, B., Laguna, M.: Fine-tuning of algorithms using fractional experimental designs and local search. Operations Research 54(1), 99–114 (2006)
2. Barr, R., Golden, B., Kelly, J., Rescende, M., Stewart, W.: Designing and reporting on computational experiments with heuristic methods. Journal of Heuristics 1(1), 9–32 (1995)
3. Birattari, M., Stützle, T., Paquete, L., Varrentrapp, K.: A racing algorithm for configuring metaheuristics. In: Langdon, W.B. (ed.) Proceedings of the Genetic and Evolutionary Computation Conference, pp. 11–18. Morgan Kaufmann, San Francisco, CA, USA (2002)
4. Birattari, M.: The Problem of Tuning Metaheuristics as Seen from a Machine Learning Perspective. PhD thesis, Université Libre de Bruxelles, Brussels, Belgium (2004)
5. Becker, S., Gottlieb, J., Stützle, T.: Applications of racing algorithms: An industrial perspective. In: Talbi, E.-G., Liardet, P., Collet, P., Lutton, E., Schoenauer, M. (eds.) EA 2005. LNCS, vol. 3871, pp. 271–283. Springer, Heidelberg (2006)

6. Chiarandini, M., Birattari, M., Socha, K., Rossi-Doria, O.: An effective hybrid algorithm for university course timetabling. Journal of Scheduling 9(5), 403–432 (2006)
7. Pellegrini, P., Birattari, M.: The relevance of tuning the parameters of metaheuristics. A case study: The vehicle routing problem with stochastic demand. Technical Report TR/IRIDIA/2006-008, IRIDIA, Université Libre de Bruxelles, Brussels, Belgium (2006)
8. Zlochin, M., Birattari, M., Meuleau, N., Dorigo, M.: Model-based search for combinatorial optimization: A critical survey. Annals of Operations Research 131, 373–395 (2004)
9. Maron, O., Moore, A.: Hoeffding races: Accelerating model selection search for classification and function approximation. In: Cowan, J.D., Tesauro, G., Alspector, J. (eds.) NIPS, vol. 6, pp. 59–66. Morgan Kaufmann, San Francisco, CA, USA (1994)
10. Moore, A., Lee, M.: Efficient algorithms for minimizing cross validation error. In: Proceedings of the Eleventh International Conference on Machine Learning, pp. 190–198. Morgan Kaufmann, San Francisco, CA, USA (1994)
11. Conover, W.J.: Practical Nonparametric Statistics, 3rd edn. John Wiley & Sons, New York,USA (1999)
12. Hoos, H., Stützle, T.: Stochastic Local Search: Foundations and Applications. Morgan Kaufmann, San Francisco, CA, USA (2005)
13. Stützle, T., Hoos, H.: $\mathcal{MAX}$–$\mathcal{MIN}$ Ant System. Future Generation Computer System 16(8), 889–914 (2000)
14. Birattari, M., Balaprakash, P., Stützle, T., Dorigo, M.: Estimation-based local search for stochastic combinatorial optimization. Technical Report TR/IRIDIA/2007-003, IRIDIA, Université Libre de Bruxelles, Brussels, Belgium (2007)
15. Birattari, M.: The race package for R. Racing methods for the selection of the best. Technical Report TR/IRIDIA/2003-37, IRIDIA, Université Libre de Bruxelles, Brussels, Belgium (2003) Package available at: http://cran.r-project.org/src/contrib/Descriptions/race.html
16. Dorigo, M., Stützle, T.: Ant Colony Optimization. MIT Press, Cambridge, MA (2004)
17. Johnson, D.S., McGeoch, L.A., Rego, C., Glover, F.: 8th DIMACS implementation challenge (2001)
18. Kohavi, R., John, G.: Automatic parameter selection by minimizing estimated error. In: Prieditis, A., Russell, S. (eds.) Proceedings of the Twelfth International Conference on Machine Learning, pp. 304–312 (1995)
19. Boyan, J., Moore, A.: Using prediction to improve combinatorial optimization search. In: Sixth International Workshop on Artificial Intelligence and Statistics (1997)
20. Audet, C., Orban, D.: Finding optimal algorithmic parameters using the mesh adaptive direct search algorithm. SIAM Journal on Optimization 17(3), 642–664 (2006)
21. Beielstein, T., Parsopoulos, K., Vrahatis, M.: Tuning PSO parameters through sensitivity analysis. Technical report, Collaborative Research Center 531 Computational Intelligence CI-124/02 (2002)
22. Hutter, F., Hoos, H., Stützle, T.: Automatic algorithm configuration based on local search. In: AAAI-2007, AAAI Press, Menlo Park, CA, USA (2007)

# Using Branch & Bound Concepts in Construction-Based Metaheuristics: Exploiting the Dual Problem Knowledge*

Christian Blum[1] and Monaldo Mastrolilli[2]

[1] ALBCOM research group
Universitat Politècnica de Catalunya, Barcelona, Spain
cblum@lsi.upc.edu
[2] IDSIA, Manno, Switzerland
monaldo@idsia.ch

**Abstract.** In recent years it has been shown by means of practical applications that the incorporation of branch & bound concepts within construction-based metaheuristics can be very useful. In this paper, we attempt to give an explanation of why this type of hybridization works. First, we introduce the concepts of primal and dual problem knowledge, and we show that metaheuristics only exploit the primal problem knowledge. In contrast, hybrid metaheuristic that include branch & bound concepts exploit both the primal and the dual problem knowledge. After giving a survey of these techniques, we conclude the paper with an application example that concerns the longest common subsequence problem.

## 1   Introduction

One of the basic ingredients of an optimization technique is a mechanism for exploring the search space. An important class of algorithms tackles an optimization problem by exploring the search space in form a of a tree, the so-called *search tree*. The search tree is generally defined by an underlying solution construction mechanism. Each path from the root node of the search tree to one of the leafs corresponds to the process of constructing a candidate solution. Inner nodes of the tree can be seen as partial solutions. The process of moving from an inner node to one of its child nodes is called a solution construction step, or extension of a partial solution.

The above mentioned class of algorithms comprises approximate as well as complete techniques. Examples of approximate methods are ant colony optimization (ACO) [6] and greedy randomized adaptive search procedures (GRASP) [7].

---

* Christian Blum acknowledges support from the Spanish CICYT project OPLINK (grant TIN-2005-08818-C04-01), and from the *Ramón y Cajal* program of the Spanish Ministry of Science and Technology of which he is a research fellow. Monaldo Mastrolilli acknowledges support from the Swiss National Science Foundation project 200021-104017/1, *Power Aware Computing*, and by the Swiss National Science Foundation project 200021-100539/1, *Approximation Algorithms for Machine scheduling Through Theory and Experiments*.

T. Bartz-Beielstein et al. (Eds.): HM 2007, LNCS 4771, pp. 123–139, 2007.

They are iterative algorithms that employ repeated probabilistic solution constructions at each iteration. While ACO algorithms include a learning component, GRASP algorithms generally do not. An example of a complete method is branch & bound. An intersting heuristic version of a breadth-first branch & bound is *beam search* [15]. While branch & bound (implicitly) considers all nodes of a certain level of the search tree, beam search restricts the search to a certain number of nodes based on the bounding information.

Both types of algorithms mentioned above have advantages as well as disadvantages. While ACO and GRASP can generally find good solutions in a reasonable amount of time, they have no mechanism to avoid wasting computation time by visiting the same solution more than once. Complete techniques on the other side guarantee to find an optimal solution. However, a user might not be prepared to accept overly large running times. One relatively recent line of research deals with the incorporation of features originating from deterministic branch & bound derivatives such as beam search into construction-based metaheuristics. Examples are probabilistic beam search (PBS) [4], Beam-ACO algorithms [2,3], and approximate and non-deterministic tree search (ANTS) procedures [12,13,14].

The aims of this paper are twofold. First, we want to give a motivation of why branch & bound features should be incorporated in construction-based metaheuristics. This is done by the introduction of different types of problem knowledge in Section 3. We will show that construction-based metaheuristics and branch & bound derivatives are complementary in their way of exploiting the problem knowledge. In Section 4 we give an overview over the existing hybrid algorithms. The second aim of the paper consists in providing an application example. This is done in Section 5. Finally, in Section 6 we conclude.

## 2   A Tree Search Model

The following tree search model captures the essential elements common to all constructive procedures. In general, we are given an optimization problem $\mathcal{P}$ and an instance $x$ of $\mathcal{P}$. Typically, the search space $S_x$ is exponentially large in the size of the input $x$. Without loss of generality we intend to maximize the objective function $f : S_x \mapsto \mathbb{R}^+$. The optimization goal is to find a solution $y \in S_x$ to $x$ with $f(y)$ as great as possible. Assume that each element $y \in S_x$ can be viewed as a composition of $l_{y,x} \in \mathbb{N}$ elements from a set $\Sigma$. From this point of view, $S_x$ can be seen as a set of strings over an alphabet $\Sigma$. Any element $y \in S_x$ can be constructed by concatenating $l_{y,x}$ elements of $\Sigma$.

The following method for constructing elements of $S_x$ is instructive: A solution construction starts with the empty string $\epsilon$. The construction process consists of a sequence of construction steps. At each construction step, we select an element of $\Sigma$ and append it to the current string $t$. The solution construction may end for two reasons. First, it may end in case $t$ has no feasible extensions. This happens when $t$ is already a complete solution, or when no solution of $S_x$ has prefix $t$. Second, a solution construction ends in case of available upper bound

---

**Algorithm 1.** Solution construction: $SC(\hat{f})$

---

1: **input:** the best known objective function value $\hat{f}$ (might be 0)
2: **initialization:** $v := v_0$
3: **while** $|\mathcal{C}(v)| > 0$ **and** $v \neq$ NULL **do**
4:     $w := \mathsf{ChooseFrom}(\mathcal{C}(v))$
5:     **if** $UB(w) > \hat{f}$ **then** $v :=$ NULL **else** $v := w$ **endif**
6: **end while**
7: **output:** $v$ (which is either a complete solution, or NULL)

---

information that indicates that each solution with prefix $t$ is worse than any solution that is already known. Henceforth we denote the upper bound value of a partial solution $t$ by $UB(t)$.

The application of such an algorithm can be equivalently described as a walk from the root $v_0$ of the corresponding search tree to a node at level $l_{y,x}$. The search tree has nodes for all $y \in S_x$ and for all prefixes of elements of $S_x$. The root of the tree is the empty string, that is, $v_0$ corresponds to $\epsilon$. There is a directed arc from node $v$ to node $w$ if $w$ can be obtained by appending an element of $\Sigma$ to $v$. Note that henceforth we identify a node $v$ of the search tree with its corresponding string $t$. We will use both notations interchangably. The set of nodes that can be reached from a node $v$ via directed arcs are called the children of $v$, denoted by $\mathcal{C}(v)$. Note, that the nodes at level $i$ correspond to strings of length $i$. If $w$ is a node corresponding to a string of length $l > 0$ then the length $l - 1$ prefix $v$ of $w$ is also a node, called the father of $w$ denoted by $\mathcal{F}(w)$. Thus, every $y \in S_x$ corresponds to exactly one path of length $l_{y,x}$ from the root node of the search tree to a specific leaf. The above described solution construction process is pseudo-coded in Algorithm 1. In the following we assume function $\mathsf{ChooseFrom}(\mathcal{C}(v))$ of this algorithm to be implemented as a probabilistic choice function.

## 3   Primal and Dual Problem Knowledge

The analysis provided in the following assumes that there is a unique optimal solution, represented by leaf node $v_d$ of the search tree, also refered to as the target node. Let us assume that—without loss of generality—the target node $v_d$ is at the maximum level $d \geq 1$ of the search tree. A probabilistic constructive optimization algorithm is said to be *successful*, if it can find the target node $v_d$ with high probability.

In the following let us examine the success probability of repeated applications of Algorithm 1 in which function $\mathsf{ChooseFrom}(\mathcal{C}(v))$ is implemented as a probabilisitc choice function. Such solution constructions are employed, for example, within the ACO metaheuristic. The value of the input $\hat{f}$ is not important for the following analysis. Given any node $v_i$ at level $i$ of the search tree, let $\mathbf{p}(v_i)$ be the probability that a solution construction process includes node

$v_i$. Note that there is a single path from $v_0$, the root node, to $v_i$. We denote the corresponding sequence of nodes by $(v_0, v_1, v_2, ..., v_i)$. Clearly, $\mathbf{p}(v_0) = 1$ and $\mathbf{p}(v_i) = \prod_{j=0}^{i-1} \mathbf{p}(v_{j+1}|v_j)$. Hereby, the probabilities $\mathbf{p}(v_{j+1}|v_j)$ are called the transition probabilities. Let $Success(\rho)$ denote the event of finding the target node $v_d$ within $\rho$ applications of Algorithm 1. Note that the probability of $Success(\rho)$ is equal to $1 - (1 - \mathbf{p}(v_d))^\rho$, and it is easy to check that the following inequalities hold:

$$1 - e^{-\rho \mathbf{p}(v_d)} \leq 1 - (1 - \mathbf{p}(v_d))^\rho \leq \rho \mathbf{p}(v_d) \tag{1}$$

By (1), it immediately follows that the chance of finding node $v_d$ is *large* if and only if $\rho \mathbf{p}(v_d)$ is *large*, namely as soon as

$$\rho = O\left(1/\mathbf{p}(v_d)\right) \ . \tag{2}$$

In the following, we will not assume anything about the exact form of the given probability distribution. However, let us assume that the transition probabilities are heuristically related to the *attractiveness* of child nodes. In other words, we assume that in a case in which a node $v$ has two children, say $w$ and $q$, and $w$ is known (or believed) to be *more promising*, then $\mathbf{p}(w|v) > \mathbf{p}(q|v)$. This can be achieved, for example, by defining the transition probabilities proportional to the weights assigned by greedy functions.

Clearly, the probability distribution reflects the available knowledge on the problem, and it is composed of two types of knowledge. If the probability $\mathbf{p}(v_d)$ of reaching the target node $v_d$ is "high", then we have a "good" problem knowledge. Let us call the knowledge that is responsible for the value of $\mathbf{p}(v_d)$ the **primal problem knowledge** (or just primal knowledge). From the dual point of view, we still have a "good" knowledge of the problem if for "most" of the wrong nodes (i.e. those that are not on the path from $v_0$ to $v_d$) the probability that they are reached is "low". We call this knowledge the **dual problem knowledge** (or just dual knowledge). Note that the quality of the dual knowledge grows with the value $\hat{f}$ that is provided as input to Algorithm 1. This means, the better the solution that we already know, the higher is the quality of the dual knowledge. Observe that the two types of problem knowledge outlined above are complementary, but not the same. Let us make an example to clarify these two concepts. Consider the search tree of Figure 1, where the target node is $v_5$. Let us analyze two different probability distributions:

**Case (a).** For each $v$ and $w \in C(v)$ let $\mathbf{p}(w|v) = 0.5$. Moreover, let us assume that no upper bound information is available. This means that each solution construction is performed until a leaf node is reached. When probabilistically constructing a solution the probability of each child is therefore the same at each construction step.

**Case (b).** In general, the transition probabilities are defined as in case (a), with one exception. Let us assume that the available upper bound indicates that the subtrees rooted in the black nodes do not contain any better solutions

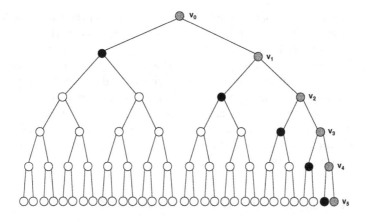

**Fig. 1.** Example of a search tree. $v_5$ is the unique optimal solution

than the ones we already know, that is, $\text{UB}(v) \leq \hat{f}$, where $v$ is a black node. Accordingly, the white children of the black nodes have probability 0 to be reached.

Note that in both cases the primal knowledge is "scarce", since the probability that the target node $v_d$ is reached by a probabilistic solution construction decreases exponentially with $d$, that is, $\mathbf{p}(v_d) = 2^{-d}$. However, in **case (b)** the dual knowledge is "excellent", since for most of the wrong nodes (i.e. the white nodes), the probability that any of them is reached is zero. Viceversa, in **case (a)** the dual knowledge is "scarce", because there is a relatively "high" probability that a white node is reached.

By using the intuition given by the provided example, let us try to better quantify the quality of the available problem knowledge. Let $V_i$ be the set of nodes at level $i$, and let

$$\ell(i) = \sum_{v \in V_i} \mathbf{p}(v), \qquad i = 1, \ldots, d . \qquad (3)$$

Note that $\ell(i)$ is equal to the probability that the solution construction process reaches level $i$ of the search tree. Observe that the use of the upper bound information makes the probabilities $\ell(i)$ smaller than one. **Case (b)** was obtained from **case (a)** by decreasing $\ell(i)$ (for $i = 1, \ldots, d$) down to $2^{i-1}$ (and without changing the probability $\mathbf{p}(v_i)$ of reaching the ancestor $v_i$ of the target node at level $i$), whereas in **case (a)** it holds that $\ell(i) = 1$ (for $i = 1, \ldots, d$). In general, good dual knowledge is supposed to decrease $\ell(i)$ without decreasing the probability of reaching the ancestor $v_i$ of the target node $v_d$. This discussion may suggest that a characterization of the available problem knowledge can be given by the following *knowledge ratio*:

$$K_{v_d} = \min_{1 \leq i \leq d} \frac{\mathbf{p}(v_i)}{\ell(i)} \qquad (4)$$

The larger this ratio the better the knowledge we have on the target node $v_d$. In case (a) it is $K_{v_d} = 1/2^d$, whereas the knowledge ratio of case (b) is $K_{v_d} = 1/2$, which is exponentially larger.

Finally, it is important to observe that the way of (repeatedly) constructing solutions in a probabilistic way (as in Algorithm 1) does not exploit the dual problem knowledge. For example in case (b), although the available knowledge is "excellent", the target node $v_d$ is found after an expected number of runs that is proportional to $1/Pr[v_d] = 2^d$ (see Equation (2)), which is the same as in case (a). In other words, the number of necessary probabilistic solution constructions only depends on the primal knowledge.

## 4   How to Exploit the Dual Knowledge?

The problem of Algorithm 1 is clearly the following one: When encountering a partial solution whose upper bound is less or equal to the value of the best solution found so far, the construction process is aborted, and the computation time invested in this construction is lost. Generally, this situation may occur very often. In fact, the probability for the abortion of a solution construction is $1 - \mathbf{p}(v_d)$ in the example outlined in the previous section, which is quite high.

In the following let us examine a first simple extension of Algorithm 1. The corresponding algorithm—henceforth denoted by $\mathrm{PSC}(\alpha, \hat{f})$—is pseudo-coded in Algorithm 2. Hereby, $\alpha$ denotes the maxiumum number of allowed extensions of partial solutions at each construction step; in other words, $\alpha$ is the maximum number of solutions to be constructed in parallel. We use the following additional notation: For any given set $S$ of search tree nodes let $\mathcal{C}(S)$ be the set of children of the nodes in $S$. Morever, $B_i$ denotes the set of reached nodes of tree level $i$. Recall that the root node $v_0$ is the only node at level 0.

The algorithm works as follows. Given the selected nodes $B_i$ of level $i$ (with $|B_i| \leq \alpha$)), the algorithm probabilistically chooses at most $\alpha$ solutions from $C := \mathcal{C}(B_i)$, the children of the nodes in $B_i$. The probabilistic choice of a child is performed in function ChooseFrom($C$) proportionally to the following probabilities:

$$\mathbf{p}(w|C) = \frac{\mathbf{p}(w|\mathcal{F}(w))}{\sum\limits_{v \in C} \mathbf{p}(v|\mathcal{F}(v))} \quad , \forall\, w \in C \tag{5}$$

Remember that $\mathcal{F}(w)$ denotes the father of node $w$. After chosing a node $w$ it is first checked if $w$ is a complete solution, or not. In case it is not a complete solution, it is checked if the available bounding information allows the further extension of this partial solution, in which case $w$ is added to $B_{i+1}$. However, if $w$ is already a complete solution, it is checked if its value is better than the value of the best solution found so far. The algorithm returns the best solution found, in case it is better than the $\hat{f}$ value that was provided as input. Otherwise the algorithm returns NULL.

---

**Algorithm 2.** Parallel solution construction: $\text{PSC}(\alpha, \hat{f})$

---

1: **input:** $\alpha \in \mathbb{Z}^+$, the best known objective function value $\hat{f}$
2: **initialization:** $i := 0$, $B_i := \{v_0\}$, $z := \text{NULL}$
3: **while** $B_i \neq \emptyset$ **do**
4:     $B_{i+1} := \emptyset$
5:     $C := \mathcal{C}(B_i)$
6:     **for** $k = 1, \ldots, \min\{\alpha, |\mathcal{C}(B_i)|\}$ **do**
7:         $w := \text{ChooseFrom}(C)$
8:         **if** $|\mathcal{C}(w)| > 0$ **then**
9:             **if** $\text{UB}(w) > \hat{f}$ **then** $B_{i+1} := B_{i+1} \cup \{w\}$ **end if**
10:         **else**
11:             **if** $f(w) > \hat{f}$ **then** $z := w$, $\hat{f} := f(z)$ **end if**
12:         **end if**
13:         $C := C \setminus \{w\}$
14:     **end for**
15:     $i := i + 1$
16: **end while**
17: **output:** $z$ (which might be NULL)

---

Observe that when $\alpha = 1$, $\text{PSC}(\alpha, \hat{f})$ is equivalent to $\text{SC}(\hat{f})$. In contrast, when $\alpha > 1$ the algorithm constructs (maximally) $\alpha$ solutions non-independently in parallel. Concerning the example outlined in the previous section with the probability distribution as defined in **case(b)**, we can observe that algorithm $\text{PSC}(\alpha, \hat{f})$ with $\alpha > 1$ solves this problem even within one application. This indicates that algorithm $\text{PSC}(\alpha, \hat{f})$, in contrast to algorithm $\text{SC}(\hat{f})$, benefically uses the dual problem knowledge.

### 4.1   Probabilistic Beam Search

For practical optimization, algorithm $\text{PSC}(\alpha, \hat{f})$ has some drawbacks. First, in most cases algorithms for optimization are applied with the goal of finding a solution as good as possible, without having a clue beforehand about the value of good solutions. Second, the available upper bound function might not be very tight. For both reasons, solution constructions that lead to unsatisfying solutions are discarded only at very low levels of the search tree, that is, close to the leafs. Referring to the example of Section 3, this means that black nodes will only appear close to the leafs. In those cases, algorithm $\text{PSC}(\alpha, \hat{f})$ will have practically no advantage over algorithm $\text{SC}(\hat{f})$. It might even have a disadvantage due to the amount of computation time invested in choosing children from bigger sets.

The following simple extension can help in overcoming the drawbacks of algorithm $\text{PSC}(\alpha, \hat{f})$. At each algorithm iteration we allow the choice of $\mu \cdot \alpha$ nodes from $B_i$, instead of $\alpha$ nodes. $\mu \geq 1$ is a parameter of the algorithm. Moreover, after the choice of the child nodes we restrict set $B_{i+1}$ to the (maximally) $\alpha$ best solutions with respect to the upper bound information. This results in a so-called (probabilistic) beam search algorithm, henceforth denoted by $PBS(\alpha,\mu,\hat{f})$. Note that algorithm $PBS(\alpha,\mu,\hat{f})$ is a generalization of algorithm $PSC(\alpha,\hat{f})$, that is, when $\mu = 1$ both algorithms are equivalent. Algorithm $PBS(\alpha,\mu,\hat{f})$ is also a generalization of algorithm $SC(\hat{f})$, which is obtained by $\alpha = \mu = 1$.

## 4.2    Adding a Learning Component to $PBS(\alpha,\mu,\hat{f})$

In general, algorithm $PBS(\alpha,\mu,\hat{f})$ can be expected to produce good solutions if (at least) two conditions are fullfilled: Neither the greedy function nor the upper bound function are misleading. In both cases the algorithm might not be able to find solutions above a certain threshold. One possibility of avoiding this drawback is to add a learning component to algorithm $PBS(\alpha,\mu,\hat{f})$, that is, adding a mechanism that is supposed to adapt the primal knowledge, the dual knowledge, or both, over time, based on accumulated search experience.

Ant colony optimization (ACO) [6] is the most prominent construction-based metaheuristic that attempts to learn the primal problem knowledge during runtime. ACO is inspired by the foraging behavior of ant colonies. At the core of this behavior is the indirect communication between the ants by means of chemical pheromone trails, which enables them to find short paths between their nest and food sources. This characteristic of real ant colonies is exploited in ACO algorithms in order to solve, for example, combinatorial optimization problems.

In general, the ACO approach attempts to solve an optimization problem by iterating the following two steps. First, $\alpha$ candidate solutions are probabilistically constructed. Second, the constructed solutions are used to modifiy the primal problem knowledge. While standard ACO algorithms use $\alpha$ applications of algorithm $SC(\hat{f})$ at each iteration for the probabilistic construction of solutions, the idea of Beam-ACO [2,3] is to use one application of probabilistic beam search $PBS(\alpha,\mu,\hat{f})$ instead. A related ACO approach is labelled ANTS (see [12,13,14]). The characterizing feature of ANTS is the use of upper bound information for defining the primal knowledge. The latest version of ANTS [14] uses at each iteration algorithm $PSC(\alpha,\hat{f})$ to construct candidate solutions.

## 5    Example: The Longest Common Subsequence Problem

The longest common subsequence (LCS) problem is one of the classical string problems. Given a problem instance $(\mathcal{S}, \Sigma)$, where $\mathcal{S} = \{s_1, s_2, \ldots, s_n\}$ is a set of $n$ strings over a finite alphabet $\Sigma$, the problem consists in finding a longest string $t^*$ that is a subsequence of all the strings in $\mathcal{S}$. Such a string $t^*$ is called a *longest common subsequence* of the strings in $\mathcal{S}$. Note that a string $t$ is called a subsequence of a string $s$, if $t$ can be produced from $s$ by deleting characters. For example, *dga* is a subsequence of *adagtta*. If $n = 2$ the problem is polynomially solvable, for example, by dynamic programming [9]. However, when $n > 2$ the problem is in general NP-hard [11]. Traditional applications of this problem are in data compression, syntactic pattern recognition, and file comparison [1], whereas more recent applications also include computational biology [16].

**Fig. 2.** Given is the problem instance $(S = \{s_1, s_2, s_3\}, \Sigma = \{a, b, c, d\})$ where $s_1 = acbcadbbd$, $s_2 = cabdacdcd$, and $s_3 = babcddaab$. Let us assume that $t = abcd$. (a), (b), and (c) show the corresponding division of $s_i$ into $s_i^A$ and $s_i^B$, as well as the setting of the pointers $p_i$ and the next positions of the 4 letters in $s_i^B$. Note that in case a letter does not appear in $s_i^B$ (for example, letter $a$ does not appear in $s_1^B$), the corresponding pointer is set to $\infty$. For example, as letter $a$ does not appear in $s_1^B$, we set $1_a := \infty$.

## 5.1  Probabilistic Beam Search for the LCS Problem

In order to apply algorithm $\mathrm{PBS}(\alpha, \mu, \hat{f})$ to the LCS problem, we have to define the solution construction mechanism, the greedy function that defines the primal knowledge, and the upper bound function that defines the dual knowledge. We use the construction mechanism of the so-called BEST-NEXT heuristic [8,10] for our algorithm. Given a problem instance $(S, \Sigma)$, this heuristic produces a common subsequence $t$ sequentially by appending at each construction step a letter to $t$ such that $t$ maintains the property of being a common subsequence of all strings in $S$. Given a common subsequence $t$ of the strings in $S$, we explain in the following how to derive the children of $t$. For that purpose we introduce the following notations:

1. Let $s_i = s_i^A \cdot s_i^B$ be the partition of $s_i$ into substrings $s_i^A$ and $s_i^B$ such that $t$ is a subsequence of $s_i^A$ and $s_i^B$ has maximal length. Given this partition, which is well-defined, we introduce position pointers $p_i := |s_i^A|$ for $i = 1, \ldots, n$ (see Figure 2 for an example).
2. The position of the first appearance of a letter $a \in \Sigma$ in a string $s_i \in S$ after the position pointer $p_i$ is well-defined and denoted by $i_a$. In case a letter $a \in \Sigma$ does not appear in $s_i^B$, $i_a$ is set to $\infty$ (see Figure 2).
3. A letter $a \in \Sigma$ is called *dominated*, if there exists at least one letter $b \in \Sigma$ such that $i_b < i_a$ for $i = 1, \ldots, n$;
4. $\Sigma_t^{\mathrm{nd}} \subseteq \Sigma$ henceforth denotes the set of non-dominated letters of the alphabet $\Sigma$ with respect to a given $t$. Moreover, for all $a \in \Sigma_t^{\mathrm{nd}}$ it is required that $i_a < \infty$, $i = 1, \ldots, n$. Hence, we require that in each string $s_i$ a letter $a \in \Sigma_t^{\mathrm{nd}}$ appears at least once after position pointer $p_i$.

The children $\mathcal{C}(t)$ of a node $t$ are then determined as follows: $\mathcal{C}(t) := \{v = ta \mid a \in \Sigma_t^{\mathrm{nd}}\}$. The primal problem knowledge is derived from the greedy function $\eta(\cdot)$ that assigns to each child $v = ta \in \mathcal{C}(t)$ the following greedy weight:

$$\eta(v) = \min\{|s_i| - i_a \mid i = 1, \ldots, n\} \tag{6}$$

The child with the highest greedy weight is considered the *most promising one*. Instead of the greedy weights themselves, we will use the corresponding ranks.

More in detail, the child $v = ta$ with the highest greedy weight will be assigned rank 1, denoted by $r(v) = 1$, the child $w = tb$ with the second-hightest greedy weight will be assigned rank 2 (that is, $r(w) = 2$), and so on.

In the following we explain the implementation of function $\mathsf{ChooseFrom}(C)$ of algorithm $\mathrm{PBS}(\alpha,\mu,\hat{f})$. Remember that $C$ denotes the set of children obtained from the nodes that are contained in the beam $B_i$ (that is, $C := \mathcal{C}(B_i)$). For evaluating a child $v \in C$ we use the sum of the ranks of the greedy weights that correspond to the construction steps performed to construct string $v$. Let us assume that $v$ is on the $i$-th level of the search tree, and let us denote the sequence of characters that forms string $v$ by $v_1 \ldots v_i$, that is, $v = v_1 \ldots v_i$. Then,

$$\nu(v) := \sum_{j=1}^{i} r(v_1 \ldots v_j) \quad , \tag{7}$$

where $v_1 \ldots v_j$ denotes the substring of $v$ from position 1 to postion $j$. With this definition, Equation 5 can be defined for the LCS problem as follows:

$$\mathbf{p}(v|C) = \frac{\nu(v)^{-1}}{\sum_{w \in C} \nu(w)^{-1}} \quad , \forall\, v \in C \tag{8}$$

Finally, we outline the upper bound function $\mathrm{UB}(\cdot)$ that the $\mathrm{PBS}(\alpha,\mu,\hat{f})$ algorithm requires. Remember that a given subsequence $t$ splits each string $s_i \in S$ into a first part $s_i^A$ and into a second part $s_i^B$, that is, $s_i = s_i^A \cdot s_i^B$. Henceforth, $|s_i^B|_a$ denotes the number of occurences of letter $a$ in $s_i^B$ for all $a \in \Sigma$. Then,

$$\mathrm{UB}(t) := |t| + \sum_{a \in \Sigma} \min\{|s_i^B|_a \mid i = 1, \ldots, n\} \quad . \tag{9}$$

In words, for each letter $a \in \Sigma$ we take the minimum of the occurences of $a$ in $s_i^B$, $i = 1, \ldots, n$. Summing up these minima and adding the result to the length of $t$ results in the upper bound. This completes the description of the implementation of the $\mathrm{PBS}(\alpha,\mu,\hat{f})$ algorithm for the LCS problem.

In the following, we use algorithm $\mathrm{PBS}(\alpha,\mu,\hat{f})$ in two different ways: First, we use $\mathrm{PBS}(\alpha,\mu,\hat{f})$ in a multi-start fashion as shown in Algorithm 3, denoted by $\mathrm{MS\text{-}PBS}(\alpha,\mu)$. Second, we use $\mathrm{PBS}(\alpha,\mu,\hat{f})$ within a Beam-ACO algorithm as explained in the following.

## 5.2 Beam-ACO for the LCS Problem

The first step of defining a Beam-ACO approach—and, in general, any ACO algorithm—consists in the specification of the set of pheromone values $\mathcal{T}$. In the case of the LCS problem $\mathcal{T}$ contains for each position $j$ of a string $s_i \in S$ a pheromone value $0 \leq \tau_{ij} \leq 1$, that is, $\mathcal{T} = \{\tau_{ij} \mid i = 1, \ldots, n, \ j = 1, \ldots, |s_i|\}$. A value $\tau_{ij} \in \mathcal{T}$ indicates the desirability of adding the letter at position $j$ of string $i$ to a solution: the greater $\tau_{ij}$, the greater is the desirability of adding the corresponding letter. In addition to the definition of the pheromone values,

---

**Algorithm 3.** Multi-start probabilistic beam search: MS-PBS($\alpha,\mu$)

---

1: **input:** $\alpha, \mu \in \mathbb{Z}^+$
2: $z :=$ NULL
3: $\hat{f} := 0$
4: **while** CPU time limit not reached **do**
5:     $v :=$ PBS($\alpha,\mu,\hat{f}$) {see Section 4.1}
6:     **if** $v \neq$ NULL **then** $z := v$, $\hat{f} := |z|$
7: **end while**
8: **output:** $z$

---

we also introduce a solution representation that is more suitable for ACO. Any common subsequence $t$ of the strings in $S$ can be translated into an ACO-solution $T = \{T_{ij} \in \{0,1\} \mid i = 1, \ldots, n, \; j = 1, \ldots, |s_i|\}$ where $T_{ij} = 0$ if the letter at position $j$ of string $s_i$ was **not** added to $t$ during the solution construction, and $T_{ij} = 1$ otherwise. Note that the translation of $t$ into $T$ is well-defined due to the construction mechanism. For example, given solution $t = abcdd$ for the problem instance of Figure 2, the corresponding ACO-solution is $T_1 = 101101001$, $T_2 = 011001101$, and $T_3 = 011111000$, where $T_i$ refers to the sequence $T_{i1} \ldots T_{i|s_i|}$. In the following, for each given solution, the lower case notation refers to its string representation, and the upper case notation refers to its binary representation.

The particular ACO framework that we used for our algorithm is the so-called $\mathcal{MMAS}$ algorithm implemented in the hyper-cube framework (HCF); see [5]. A high level description of the algorithm is given in Algorithm 4. The data structures used, in addition to counters and to the pheromone values, are: (1) the *best-so-far* solution $T^{bs}$, i.e., the best solution generated since the start of the algorithm; (2) the *restart-best* solution $T^{rb}$, that is, the best solution generated since the last restart of the algorithm; (3) the *convergence factor* $cf$, $0 \leq cf \leq 1$, which is a measure of how far the algorithm is from convergence; and (4) the Boolean variable *bs_update*, which becomes true when the algorithm reaches convergence.

Roughly, the algorithm works as follows. First, all the variables are initialized. In particular, the pheromone values are set to their initial value 0.5. Each algorithm iteration consists of the following steps. First, algorithm PBS($\alpha,\mu,\hat{f}$) is applied with $\hat{f} = 0$ to generate a solution $T^{pbs}$. The setting of $\hat{f} = 0$ is chosen, because in ACO algorithms it is generally useful to learn also from solutions that are worse than the best solution found so far. The only change in algorithm PBS($\alpha,\mu,\hat{f}$) occurs in the definition of the choice probabilities. Instead of using Equation 8, these probabilities are now defined as follows:

$$\mathbf{p}(v = ta|C) = \frac{\left( \min_{i=1,\ldots,n} \{\tau_{ii_a}\} \cdot \nu(v)^{-1} \right)}{\sum_{w=tb \in C} \left( \min_{i=1,\ldots,n} \{\tau_{ii_b}\} \cdot \nu(w)^{-1} \right)} \quad , \forall \, v = ta \in C \qquad (10)$$

---

**Algorithm 4.** Beam-ACO for the LCS problem

---

1: **input:** $\alpha, \mu \in \mathbb{Z}^+$
2: $T^{bs} :=$ NULL, $T^{rb} :=$ NULL, $cf := 0$, $bs\_update :=$ FALSE
3: $\tau_{ij} := 0.5$, $i = 1, \ldots, n$, $j = 1, \ldots, |s_i|$
4: **while** CPU time limit not reached **do**
5:    $T^{pbs} :=$ PBS$(\alpha,\mu,0)$ {see Section 4.1}
6:    **if** $|t^{pbs}| > |t^{rb}|$ **then** $T^{rb} := T^{pbs}$
7:    **if** $|t^{pbs}| > |t^{bs}|$ **then** $T^{bs} := T^{pbs}$
8:    ApplyPheromoneUpdate($cf$,$bs\_update$,$T$,$T^{pbs}$,$T^{rb}$,$T^{bs}$)
9:    $cf :=$ ComputeConvergenceFactor($T$)
10:    **if** $cf > 0.99$ **then**
11:       **if** $bs\_update =$ TRUE **then**
12:          $\tau_{ij} := 0.5$, $i = 1, \ldots, n$, $j = 1, \ldots, |s_i|$
13:          $T^{rb} :=$ NULL
14:          $bs\_update :=$ FALSE
15:       **else**
16:          $bs\_update :=$ TRUE
17:       **end if**
18:    **end if**
19: **end while**
20: **output:** $t^{bs}$ (that is, the string version of ACO-solution $T^{bs}$)

---

Remember in this context, that $i_a$ was defined as the next position of letter $a$ after position pointer $p_i$ in string $s_i$. The intuition of choosing the minimum of the pheromone values corresponding to the next positions of a letter in the $n$ given strings is as follows: If at least one of these pheromone values is low, the corresponding letter should not yet be appended to the string, because there is another letter that should be appended first.

The second action at each iteration concerns the pheromone update conducted in the ApplyPheromoneUpdate($cf$, $bs\_update$, $T$, $T^{pbs}$, $T^{rb}$, $T^{bs}$) procedure. Third, a new value for the convergence factor $cf$ is computed. Depending on this value, as well as on the value of the Boolean variable $bs\_update$, a decision on whether to restart the algorithm or not is made. If the algorithm is restarted, all the pheromone values are reset to their initial value (that is, 0.5). The algorithm is iterated until the CPU time limit is reached. Once terminated, the algorithm returns the string version $t^{bs}$ of the best-so-far ACO-solution $T^{bs}$. In the following we describe the two remaining procedures of Algorithm 4 in more detail.

ApplyPheromoneUpdate($cf$,$bs\_update$,$T$,$T^{pbs}$,$T^{rb}$,$T^{bs}$): In general, three solutions are used for updating the pheromone values. These are the solution $T^{pbs}$ generated by the PBS algorithm, the restart-best solution $T^{rb}$, and the best-so-far solution $T^{bs}$. The influence of each solution on the pheromone update depends on the state of convergence of the algorithm as measured by the convergence factor $cf$. Each pheromone value $\tau_{ij} \in T$ is updated as follows:

$$\tau_{ij} := \tau_{ij} + \rho \cdot (\xi_{ij} - \tau_{ij}) \ , \tag{11}$$

where

$$\xi_{ij} := \kappa_{pbs} \cdot T_{ij}^{pbs} + \kappa_{rb} \cdot T_{ij}^{rb} + \kappa_{bs} \cdot T_{ij}^{bs} , \qquad (12)$$

where $\kappa_{pbs}$ is the weight (that is, the influence) of solution $T^{pbs}$, $\kappa_{rb}$ is the weight of solution $T^{rb}$, $\kappa_{bs}$ is the weight of solution $T^{bs}$, and $\kappa_{pbs} + \kappa_{rb} + \kappa_{bs} = 1$. After the pheromone update rule (Equation 11) is applied, pheromone values that exceed $\tau_{max} = 0.999$ are set back to $\tau_{max}$ (similarly for $\tau_{min} = 0.001$). This is done in order to avoid a complete convergence of the algorithm, which is a situation that should be avoided. Equation 12 allows to choose how to schedule the relative influence of the three solutions used for updating the pheromone values. For our application we used a standard update schedule as shown in Table 1.

**Table 1.** Setting of $\kappa_{pbs}$, $\kappa_{rb}$, $\kappa_{bs}$, and $\rho$ depending on the convergence factor $cf$ and the Boolean control variable $bs\_update$

|  | bs_update = FALSE | | | | bs_update = TRUE |
|---|---|---|---|---|---|
|  | $cf < 0.4$ | $cf \in [0.4, 0.6)$ | $cf \in [0.6, 0.8)$ | $cf \geq 0.8$ |  |
| $\kappa_{ib}$ | 1 | 2/3 | 1/3 | 0 | 0 |
| $\kappa_{rb}$ | 0 | 1/3 | 2/3 | 1 | 0 |
| $\kappa_{bs}$ | 0 | 0 | 0 | 0 | 1 |
| $\rho$ | 0.2 | 0.2 | 0.2 | 0.15 | 0.15 |

ComputeConvergenceFactor($\mathcal{T}$): The convergence factor $cf$, which is a function of the current pheromone values, is computed as follows:

$$cf := 2 \left( \left( \frac{\sum_{\tau_{ij} \in \mathcal{T}} \max\{\tau_{max} - \tau_{ij}, \tau_{ij} - \tau_{min}\}}{|\mathcal{T}| \cdot (\tau_{max} - \tau_{min})} \right) - 0.5 \right)$$

In this way, $cf = 0$ when the algorithm is initialized (or reset), that is, when all pheromone values are set to 0.5. On the other side, when the algorithm has converged, then $cf = 1$. In all other cases, $cf$ has a value in $(0, 1)$. This completes the description of our Beam-ACO approach for the LCS problem.

### 5.3  Experimental Results

We implemented algorithms MS-PBS($\alpha,\mu$) and Beam-ACO in ANSI C++ using GCC 3.2.2 for compiling the software. The experimental results that we outline in the following were obtained on a PC with an AMD64X2 4400 processor and 4 Gigabyte of memory. We applied algorithm MS-PBS($\alpha,\mu$) with three different settings:

1. $\alpha = \mu = 1$: The resulting algorithm corresponds to a multi-start version of algorithm SC($\hat{f}$); see Algorithm 1. In the following we refer to this algorithm as **MS-SC**.

2. $\alpha = 10$, $\mu = 1$: This setting corresponds to a multi-start version of algorithm PSC($\alpha,\hat{f}$); see Algorithm 2. We refer henceforth to this algorithm as **MS-PSC**.

3. $\alpha = 10$, $\mu > 1$: These settings generate a multi-start version of algorithm PBS($\alpha,\mu,\hat{f}$); see Section 4.1. This algorithm version is referred to simply as **MS-PBS**. Note that we made the setting of $\mu$ depended on the alphabet size, that is, the number of expected children of a partial solution.

In addition we applied Beam-ACO with $\alpha = 10$ and with the same settings for $\mu$ as chosen for MS-PBS.

For the experimentation we used a set of benchmark instances that was generated as follows. Given $h \in \{100, 200, \ldots, 1000\}$ and $\Sigma$ (where $|\Sigma| \in \{2, 4, 8, 24\}$), an instance is produced as follows. First, a string $s$ of length $h$ is produced randomly from the alphabet $\Sigma$. String $s$ is in the following called *base string*. Each instance contains 10 strings. Each of these strings is produced from the base string $s$ by traversing $s$ and by deciding for each letter with a probabilitiy of 0.1 whether to remove it, or not. Note that the 10 strings of such an instance are not necessarily of the same length. As we produced 10 instances for each combination of $h$ and $|\Sigma|$, 400 instances were generated in total. Note that the values of optimal solutions of these instances are unknown. However, a lower bound is obtained as follows. While producing the 10 strings of an instance, we record for each position of the base string $s$, whether the letter at that position was removed for the generation of at least one of the 10 strings. The number of positions in $s$ that were never removed constitutes the lower bound value henceforth denoted by $LB_I$ with respect to an instance $I$.

We applied each of the 4 algorithms exactly once for $h/10$ seconds to each problem instance. We present the results averaged over the 10 instances for each combination of $h$ (the length of the base string that was used to produce an instance), and the alphabet size $|\Sigma|$. Two measures are presented:

1. The (average) length of the solutions expressed in percent deviation from the respective lower bounds, which is computed as $((f/LB_I) - 1) \cdot 100$, where $f$ is the length of the solution achieved by the respective algorithm.

2. The computation time of the algorithms, which refers to the time the best solution was found within the given CPU time (averaged over the 10 instances of each type).

The results are shown graphically in Figure 3. The graphics on the left hand side show the algorithm performance (in percentage deviation from the lower bound), and the graphics on the right hand side show the computation times. The following observations are of interest. First, while having a comparable computation time, algorithm MS-PBS is always clearly better than algorithms MS-PSC and MS-SC. Second, algorithm Beam-ACO is consistently the best algorithm of the comparison. This shows that it can pay off adding a learning component to algorithm (MS-)PBS. The advantage of Beam-ACO over MS-PBS grows with growing alphabet size, that is, with growing problem complexity. This advantage of Beam-ACO comes with a slight increase in computational

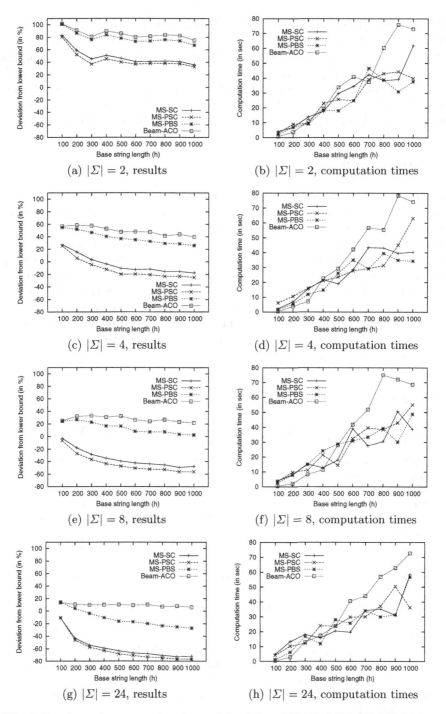

(a) $|\Sigma| = 2$, results

(b) $|\Sigma| = 2$, computation times

(c) $|\Sigma| = 4$, results

(d) $|\Sigma| = 4$, computation times

(e) $|\Sigma| = 8$, results

(f) $|\Sigma| = 8$, computation times

(g) $|\Sigma| = 24$, results

(h) $|\Sigma| = 24$, computation times

**Fig. 3.** Results and computation times of algorithms MS-SC, MS-PSC, MS-PBS, and Beam-ACO

cost. However, this is natural: due to the learning component, Beam-ACO has a higher probability than MS-PBS of improving on the best solution found even at late stages of a run. Finally, a last interesting remark concerns the comparison of MS-PSC with MS-SC. Despite of the construction of solutions in parallel, MS-PSC is always slightly beaten by MS-SC. This is due to fact that the used upper bound function is not tight at all. Hence, constructing solutions in parallel in the way of algorithm (MS-)PSC is rather a waste of computation time.

## 6   Conclusions

In this paper we have first given a motivation for the use of branch & bound concepts within construction based metaheuristics such as ant colony optimization. In this context we have introduced the definitions of primal and dual problem knowledge. Then we have shown that a certain way of using branch & bound concepts leads to the parallel and non-independent construction of solutions. A prominent example is probabilistic beam search. An extension of probabilistic beam search is obtained by adding a learning component that modifies the primal problem knowledge over time. Finally, we have implemented these algorithms on the example of the longest common subsequence problems. The results confirmed our earlier considerations on the advantage of the proposed hybrids over standard algorithms.

## References

1. Aho, A., Hopcroft, J., Ullman, J.: Data structures and algorithms. Addison-Wesley, Reading (1983)
2. Blum, C.: Beam-ACO–hybridizing ant colony optimization with beam search: an application to open shop scheduling. Computers and Operations Research 32, 1565–1591 (2005)
3. Blum, C., Bautista, J., Pereira, J.: Beam-ACO applied to assembly line balancing. In: Dorigo, M., Di Caro, G.A., Sampels, M. (eds.) ANTS 2002. LNCS, vol. 2463, pp. 14–27. Springer, Heidelberg (2002)
4. Blum, C., Cotta, C., Fernández, A.J., Gallardo, J.E.: A probabilistic beam search algorithm for the shortest common supersequence problem. In: Cotta, C., van Hemert, J.I., (eds.) EvoCOP 2007. LNCS, vol. 4446, pp. 36–47. Springer, Berlin (2007)
5. Blum, C., Dorigo, M.: The hyper-cube framework for ant colony optimization. IEEE Transactions on Systems, Man, and Cybernetics – Part B 34(2), 1161–1172 (2004)
6. Dorigo, M., Stuetzle, T.: Ant Colony Optimization. MIT Press, Cambridge (2004)
7. Feo, T.A., Resende, M.G.C.: Greedy randomized adaptive search procedures. Journal of Global Optimization 6, 109–133 (1995)
8. Fraser, C.B.: Subsequences and supersequences of strings. PhD thesis, University of Glasgow (1995)
9. Gusfield, D.: Algorithms on Strings, Trees, and Sequences. In: Computer Science and Computational Biology., Cambridge University Press, Cambridge (1997)

10. Huang, K., Yang, C., Tseng, K.: Fast algorithms for finding the common subsequences of multiple sequences. In: Proceedings of the International Computer Symposium, pp. 1006–1011. IEEE Computer Society Press, Los Alamitos (2004)
11. Maier, D.: The complexity of some problems on subsequences and supersequences. Journal of the ACM 25, 322–336 (1978)
12. Maniezzo, V.: Exact and Approximate Nondeterministic Tree-Search Procedures for the Quadratic Assignment Problem. INFORMS Journal on Computing 11(4), 358–369 (1999)
13. Maniezzo, V., Carbonaro, A.: An ANTS heuristic for the frequency assignment problem. Future Generation Computer Systems 16, 927–935 (2000)
14. Maniezzo, V., Milandri, M.: An ant-based framework for very strongly constrained problems. In: Dorigo, M., Di Caro, G.A., Sampels, M. (eds.) Ant Algorithms. LNCS, vol. 2463, pp. 222–227. Springer, Heidelberg (2002)
15. Ow, P.S., Morton, T.E.: Filtered beam search in scheduling. International Journal of Production Research 26, 297–307 (1988)
16. Smith, T., Waterman, M.: Identification of common molecular subsequences. Journal of Molecular Biology 147(1), 195–197 (1981)

# Gradient-Based/Evolutionary Relay Hybrid for Computing Pareto Front Approximations Maximizing the S-Metric

Michael Emmerich[1], André Deutz[1], and Nicola Beume[2]

[1]University of Leiden, Leiden Institute for Advanced Computer Science,
2333 CA Leiden, The Netherlands
http://natcomp.liacs.nl/
[2]University of Dortmund, Chair of Algorithm Engineering,
44221 Dortmund, Germany
{emmerich,deutz}@liacs.nl, nicola.beume@udo.edu,
http://Ls11-www.cs.uni-dortmund.de

**Abstract.** The problem of computing a good approximation set of the Pareto front of a multiobjective optimization problem can be recasted as the maximization of its S-metric value, which measures the dominated hypervolume. In this way, the S-metric has recently been applied in a variety of metaheuristics. In this work, a novel high-precision method for computing approximation sets of a Pareto front with maximal S-Metric is proposed as a high-level relay hybrid of an evolutionary algorithm and a gradient method, both guided by the S-metric. First, an evolutionary multiobjective optimizer moves the initial population close to the Pareto front. The gradient-based method takes this population as its starting point for computing a local maximal approximation set with respect to the S-metric. Thereby, the population is moved according to the gradient of the S-metric.

This paper introduces expressions for computing the gradient of a set of points with respect to its S-metric on basis of the gradients of the objective functions. It discusses singularities where the gradient is vanishing or differentiability is one sided. To circumvent the problem of vanishing gradient components of the S-metric for dominated points in the population a penalty approach is introduced.

In order to test the new hybrid algorithm, we compute the precise maximizer of the S-metric for a generalized Schaffer problem and show, empirically, that the relay hybrid strategy linearly converges to the precise optimum. In addition we provide first case studies of the hybrid method on complicated benchmark problems.

## 1 Introduction and Mathematical Preliminaries

In multiobjective optimization, a solution has to fulfill several objectives in the best possible way. Maximization problems can be reformulated as minimization problems, thus, without loss of generality, we can restrict our attention to those. Formally, the problem reads as follows:

$$\mathbf{f} = (f_1, \ldots, f_m)^T, \quad f_1(\mathbf{x}) \to \min, \ldots, f_m(\mathbf{x}) \to \min, \quad \mathbf{x} \in \mathcal{X}. \tag{1}$$

T. Bartz-Beielstein et al. (Eds.): HM 2007, LNCS 4771, pp. 140–156, 2007.
© Springer-Verlag Berlin Heidelberg 2007

The domain $\mathcal{X}$ is called the *decision space* or *search space* and contains all feasible solutions, and the co-domain $\mathcal{Y}$ of all $m$ objectives is called *objective space*. Here we assume continuous functions, so $\mathcal{X} \subseteq \mathbb{R}^d$ and $\mathcal{Y} \subseteq \mathbb{R}^m$.

Since the objectives are typically conflicting, there is no single best solution and the aim is to generate sets of good compromise solutions. These solutions are suggestions to the decision maker who finally chooses one for realization.

A partial order holds among the points, which is defined in the objective space and is transferred to the preimages in the search space. A point $\mathbf{x}$ is said to (weakly) dominate a point $\mathbf{x}'$ ($\mathbf{x} \prec \mathbf{x}'$), iff $\mathbf{f}(\mathbf{x}) \neq \mathbf{f}(\mathbf{x}')$ and $\forall i \in \{1, \ldots, m\}$ : $f_i(\mathbf{x}) \leq f_i(\mathbf{x}')$. A point $\mathbf{x}$ strictly dominates a point $\mathbf{x}'$ ($\mathbf{x} < \mathbf{x}'$), iff $\forall i = 1, \ldots, m : f_i(\mathbf{x}) < f_i(\mathbf{x}')$. The points that are minimal with respect to the partial order $\prec$ within a set are called non-dominated. The non-dominated points within the whole search space are called *efficient set or Pareto set* $\mathcal{X}_E$ and the set of their corresponding images is called *Pareto front* $\mathcal{Y}_N$.

Since continuous problems cannot be expected to be solved optimally, a good approximation of the Pareto front is aspired. Two sets are already incomparable, if one set contains a point that is incomparable to each point of the other set. Thus, a qualitative ranking is mostly impossible. Instead, auxiliary demands which suggest high quality are formulated for sets, such as: (1) many non-dominated points, (2) closeness to Pareto front, and (3) well-distributed along the Pareto front. From our point of view the term well-distributed means to have a regular spacing between points in regions with similar trade-off and a higher concentration of points in regions with a more balanced trade-off among the objectives.

Among the developed quality measures, the *S-metric* or *dominated hypervolume* by Zitzler and Thiele [1] is of utmost importance. It is defined as

$$\mathcal{S}(X) = \text{Lebesgue}\{\mathbf{y} \mid \exists \mathbf{y}^{(i)} : \mathbf{y}^{(i)} \prec \mathbf{y} \wedge \mathbf{y} \prec \mathbf{y}^{ref}\}, \tag{2}$$

where $\mathbf{y}^{(i)} = \mathbf{f}(\mathbf{x}^{(i)})$ are the image points of the set $X \subseteq \mathcal{X}$ under $\mathbf{f}$, and X is an approximation of the Pareto set. The reference point $\mathbf{y}^{ref}$ confines the dominated hypervolume. Note that the same definition can be used to define the S-metric for subsets of $\mathbb{R}^m$ directly. It is an alleged drawback that the reference point influences the absolute value of the metric. However, in practical settings it is often possible to state bounds for the objective function values and thus the reference point can be chosen as that upper bound vector. In addition, recent results on generalizations of the S-metric show, that the distribution of points on the Pareto front can be influenced by weighting parameters according to the user's preferences [2].

The maximal S-metric value is achieved by the Pareto front. For compact image sets of $\mathbf{f}$ and appropriately chosen reference points the maximization of the S-metric for a given number of points always results in a non-dominated set of solutions. Further properties of the S-metric were studied by Fleischer [3] and Zitzler et al. [4].

The maximization of the S-metric receives increasingly more attention as a solution principle for approximating Pareto fronts by means of a well-distributed non-dominated set. Recently, the S-metric has been used as a single-objective substitute function to guide the process of multiobjective optimizers. Accordingly, the problem of finding a good approximation of the Pareto front of the original multiobjective optimization problem can be re-stated as:

$$\mathcal{S}(X) \rightarrow \max, X \subseteq_\mu \mathcal{X} \tag{3}$$

where $X \subseteq_\mu \mathcal{X}$ means that $X$ is a set of at most $\mu$ elements from $\mathcal{X}$.

Recent work proposed methods for S-metric maximization that are based on simulated annealing, particle swarms, and evolutionary algorithms. *Evolutionary multiobjective optimization algorithms (EMOA, MOEA)* [5,6] established as efficient and robust optimizers and modern EMOA like IBEA [7], ESP [8], and SMS-EMOA successfully apply an S-metric based function to evaluate and select promising solutions, or use it for archiving [9]. The SMS-EMOA by Emmerich et al. [10] uses the S-metric in the selection method of a steady-state EMOA. It has been tested extensively on benchmarks and real-world applications, receiving results competitive or better than state-of-the-art methods in the field. In this paper we continue in the same spirit, and derive a gradient based method for solving multiobjective problems.

In this work the gradient of the S-metric at a point, representing an approximation set, is introduced to solve the optimization problem of positioning the given $\mu$ points of the set such that the S-metric value of the set is maximized. Using this gradient, we apply a simple steepest ascent method. We propose a hybridization of the gradient method with SMS-EMOA as a high-level relay (cf. Talbi [11]), meaning that autonomous algorithms are executed sequentially. The gradient method is applied after SMS-EMOA to locally optimize its final population. Thus, we combine efficient local optimization based on a new gradient-based method, with more exhaustive global optimization techniques.

As opposed to previous work on gradient based multiobjective optimization (e.g. [12,13,14,15,16]) this approach does not use gradients to improve points of the population independent of each other but, by aiming at improving the S-metric (that considers the population as one aggregate), it looks at the distribution of the entire population.

The paper is structured as follows. In Section 2 expressions of the gradient of the S-metric are derived and discussed. In Section 3 the maximal S-metric is determined analytically to verify the gradient formulation. Section 4 introduces a steepest descent gradient method for S-metric maximization. Afterward, the hybridization of this method with the evolutionary algorithm SMS-EMOA is proposed and studied on multimodal test problems (Section 5). The new methods form starting points for further studies. A summary of the results and discussion of open questions is provided in Section 6.

## 2   Gradient of the S-Metric

In this section we discuss expressions for gradient computation with respect to the S-metric and discuss its differentiability properties.

## 2.1  Mathematical Notation

In order to compute gradients of the S-metric, we represent a population $P$ of size $\mu$, $P \subseteq_\mu \mathcal{X}$, as a vector of length $\mu \cdot d$:

$$\mathbf{p} = (x_1^{(1)}, \ldots, x_d^{(1)}, \ldots, x_1^{(\mu)}, \ldots, x_d^{(\mu)})^\top = (p_1, \ldots, p_{\mu \cdot d})^\top.$$

For notational convenience we introduce *blocks* of a $\mu d$-vector as

$$\Pi(i, \mathbf{p}) = (x_1^{(i)}, \ldots, x_d^{(i)}) = (p_{(i-1) \cdot d + 1} \cdots p_{i \cdot d}).$$

The mapping from $\mu d$-vectors to populations is defined as:

$$\Psi(\mathbf{p}) = \{ (x_1^{(i)}, \ldots, x_d^{(i)})^\top \mid i \in \{1, \ldots, \mu\} \}. \tag{4}$$

Different $\mu d$-vectors may represent the same population (but not vice-versa). Every non-empty population $P \subseteq_\mu \mathcal{X}$ is represented by at least one tuple of the form above.

For optimization purposes it is sufficient to work with $\mu d$-vectors. This holds, because the set of global optima of the problem

$$\mathcal{S}(\Psi(\mathbf{p})) \to \max, \text{ subject to } \Psi(\mathbf{p}) \subseteq_\mu \mathcal{X}, \ \mathbf{p} \in \mathbb{R}^{\mu d} \tag{5}$$

can be mapped to the set of global optima of the original problem (Eq. 3) via $\Psi$. Note that for $\mathcal{X} = \mathbb{R}^d$ the constraint $\Psi(\mathbf{p}) \subseteq_\mu \mathcal{X}$ is trivially fulfilled. Moreover, the number of local optima of the new problem is usually increased, as different $\mu d$-vectors may give rise to the same population. Given one $\mu d$-vector, all equivalent representations can be obtained by permuting its blocks.

## 2.2  Definition and Analytical Calculation of S-Metric's Gradient

A general definition of the gradient for the space of $\mu d$-vectors is

$$\nabla_{\mathbf{p}} \mathcal{S} = (\frac{\partial \mathcal{S}}{\partial p_1}, \ldots, \frac{\partial \mathcal{S}}{\partial p_{\mu \cdot d}})^\top \tag{6}$$

In order to express the gradient of the S-metric in terms of the gradients of the objective functions the following structure of the composition of mappings is applied:

$$\mathbb{R}^{\mu \cdot d} \quad \underbrace{\xrightarrow{\mathbf{F}}}_{\text{decision to objective space}} \quad \mathbb{R}^{\mu \cdot m} \quad \underbrace{\xrightarrow{\mathcal{S}}}_{\text{objective space to S-metric}} \quad \mathbb{R}^+. \tag{7}$$

where $\mathbf{F}$ is defined by using the objective functions $\mathbf{f} = (f_1, f_2, \cdots, f_m)^\top$ so that $\mathbf{F}(\mathbf{x}^{(1)}, \cdots, \mathbf{x}^{(\mu)}) = (\mathbf{f}(\mathbf{x}^{(1)}), \mathbf{f}(\mathbf{x}^{(2)}), \cdots, \mathbf{f}(\mathbf{x}^{(\mu)}))^\top$ with the functions $f_i$ defined as above and $\mathcal{S}$ as the S-metric function.

The S-metric is defined on sets of points (Eq. 2), but for notational convenience, we also apply it directly to vectors which can be interpreted as sets according to the mapping $\Psi$ (Eq. 5). Using the chain rule the gradient can be rewritten as follows. Let $\mathbf{x}^{(1)}, \mathbf{x}^{(2)}, \cdots, \mathbf{x}^{(\mu)}$ be $\mu$ points in the decision space, then $\nabla \mathcal{S}(\mathbf{p})$ can be written as:

$$
\mathcal{S}' \text{ at} \begin{pmatrix} \mathbf{f}(\mathbf{x}^{(1)}) \\ \mathbf{f}(\mathbf{x}^{(2)}) \\ \cdots \\ \mathbf{f}(\mathbf{x}^{(\mu)}) \end{pmatrix} \circ \begin{pmatrix} \mathbf{f}' \text{ at } \mathbf{x}^{(1)} & 0 & 0 \cdots & 0 \\ 0 & \mathbf{f}' \text{ at } \mathbf{x}^{(2)} & 0 \cdots & 0 \\ \vdots & \vdots & \vdots \cdots & \vdots \\ 0 & 0 & 0\ 0 & \mathbf{f}' \text{ at } \mathbf{x}^{(\mu)} \end{pmatrix} \tag{8}
$$

The top level structure of the matrix associated to the linear mapping $\mathbf{F}'$ is a diagonal matrix of size $\mu$ whose diagonal elements are matrices of size $m \times d$ associated to the linear maps $\mathbf{f}'$ at $\mathbf{x}^{(j)}$, where $j = 1, 2, \cdots, \mu$ and each of the off-diagonal elements is the zero matrix of size $m \times d$ as well.

A more detailed description of this matrix is given as:

$$
\underbrace{\begin{pmatrix} \frac{\partial \mathcal{S}}{\partial y_1^{(1)}} \\ \vdots \\ \frac{\partial \mathcal{S}}{\partial y_m^{(1)}} \\ \vdots \\ \frac{\partial \mathcal{S}}{\partial y_1^{(\mu)}} \\ \vdots \\ \frac{\partial \mathcal{S}}{\partial y_m^{(\mu)}} \end{pmatrix}}_{\nabla \mathcal{S}(\mathbf{y}^{(1)}, \ldots, \mathbf{y}^{(\mu)})}^{\top} \cdot \underbrace{\begin{pmatrix} \frac{\partial f_1(\mathbf{x}^{(1)})}{\partial x_1^{(1)}} & \cdots & \frac{\partial f_1(\mathbf{x}^{(1)})}{\partial x_d^{(1)}} & 0 \cdots 0 & 0 & \cdots & 0 \\ \vdots & \vdots & \vdots & \vdots\ \vdots & \vdots & \vdots & \vdots \\ \frac{\partial f_m(\mathbf{x}^{(1)})}{\partial x_1^{(1)}} & \cdots & \frac{\partial f_m(\mathbf{x}^{(1)})}{\partial x_d^{(1)}} & 0 \cdots 0 & 0 & \cdots & 0 \\ 0 & \cdots & 0 & \vdots \cdots \vdots & 0 & \cdots & 0 \\ \vdots & \vdots & \vdots & \vdots\ \vdots & \vdots & \vdots & \vdots \\ 0 & \cdots & 0 & \vdots \cdots \vdots & 0 & \cdots & 0 \\ 0 & \cdots & 0 & 0 \cdots 0 & \frac{\partial f_1(\mathbf{x}^{(\mu)})}{\partial x_1^{(\mu)}} & \cdots & \frac{\partial f_1(\mathbf{x}^{(\mu)})}{\partial x_d^{(\mu)}} \\ \vdots & \vdots & \vdots & \vdots\ \vdots & \vdots & \vdots & \vdots \\ 0 & \cdots & 0 & 0 \cdots 0 & \frac{\partial f_m(\mathbf{x}^{(\mu)})}{\partial x_1^{(\mu)}} & \cdots & \frac{\partial f_m(\mathbf{x}^{(\mu)})}{\partial x_d^{(\mu)}} \end{pmatrix}}_{\mathbf{F}'(\mathbf{x}^{(1)}, \ldots, \mathbf{x}^{(\mu)})} \tag{9}
$$

Note that $\mathbf{F}'(\mathbf{x}^{(1)}, \ldots, \mathbf{x}^{(\mu)})$ depends solely on the gradient functions $\nabla f_i$ at the sites $\mathbf{x}^{(1)}, \ldots, \mathbf{x}^{(\mu)}$. Hence, if these $m \cdot \mu$ local gradients are known, the desired gradient $\nabla \mathcal{S}(\mathbf{p})$ can be computed.

The computation of $\nabla \mathcal{S}(\mathbf{y}^{(1)}, \ldots, \mathbf{y}^{(\mu)})$ is discussed next. Three cases of the set $\{\mathbf{y}^{(1)}, \ldots, \mathbf{y}^{(\mu)}\}$ need to be considered: (1) mutually non-dominated sets, (2) sets with strictly dominated points, and (3) sets with weakly dominated points.

*(1) Mutually non-dominated sets.* For $m = 1$ holds $\frac{\partial \mathcal{S}}{\partial y_1^{(i)}} = 1$, and for $m = 2$ holds (assuming vectors are sorted $\mathbf{y}^{(i)}$ in descending order of $f$):

$$
\frac{\partial \mathcal{S}}{\partial y_1^{(i)}} = y_2^{(i-1)} - y_2^{(i)} \quad \text{and} \quad \frac{\partial \mathcal{S}}{\partial y_2^{(i)}} = y_1^{(i-1)} - y_1^{(i)}, \quad i = 1, \ldots, \mu \tag{10}
$$

as illustrated in Fig. 1. Note that extremal points need special treatment, as their contribution to the gradient is influenced by the reference point. In three dimensions ($m = 3$), the computation of the partial derivative gets more tedious. The general principle is sketched in Fig. 2.

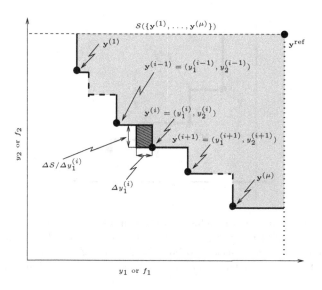

**Fig. 1.** Partial Derivative of the S-metric for $m = 2$ and non-dominated sets. The lengths of the line-segments of the attainment curve correspond to the values of the partial derivatives of $\mathcal{S}$. Only for extremal points do the values of the partial derivatives depend on the reference point.

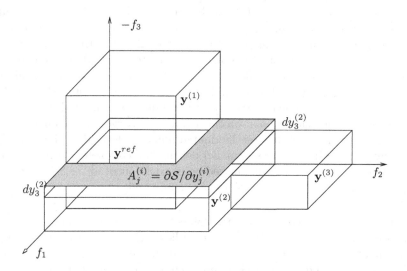

**Fig. 2.** Partial derivative for $m = 3$. By changing a point $\mathbf{y}^{(i)}$ differentially in the $j$-th coordinate direction, the hypervolume grows with the area $A_j^{(i)}$ of the 'visible' face of the exclusively contributed hypervolume of that point in the direction of the movement. Hence $A_j^{(i)}$ is the partial derivative $\partial \mathcal{S}/\partial y_j^{(i)}$.

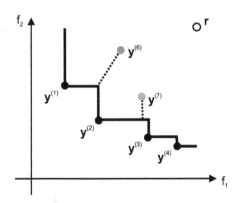

**Fig. 3.** The penalty function is defined as the sum of the euclidean distance (dashed lines) of the dominated points (gray) to the attainment curve (solid line) shaped by the non-dominated points (black) and bounded by the reference point **r**. The penalty is subtracted from the S-metric value to give an influence to the dominated points.

*(2) Sets with strictly dominated points.* The gradient equals zero in case of dominated points—provided that a slight perturbation does not make them non-dominated—since no improvement of the S-metric can be observed for any movement. Therefore, dominated points do not move during a search with gradient methods but just remain in their position. To enable an improvement of dominated points, a *penalty value* can be subtracted from the S-metric value, that is negative if and only if points are dominated and otherwise zero. For each dominated point, the minimal Euclidean distance to the attainment surface shaped by the non-dominated points is calculated (Fig. 3). The sum of these values is subtracted from the S-metric value of the whole set of points. This way, the movement of dominated points influences the improvement of the penalized S-metric and a local gradient of the dominated points is computed that points in the direction of the nearest point on the attainment curve. In a gradient descent method the movement of the non-dominated points is delayed by the dominated ones. Anyway, this drawback is a smaller deficit than completely losing the dominated points. Since any non-dominated point contributes to the S-metric value, the primary aim is to make all points non-dominated.

*(3) Sets with weakly dominated points.* Points that are dominated but not strictly dominated (we call them *weakly dominated*) lie on the attainment surface of the non-dominated points. Slight movements can make the points either remain weakly dominated, become strictly dominated or non-dominated. Thus, the gradient at these points is not continuous. The left-sided derivative $\frac{\partial^- S}{y_j^{(i)}}$ may be positive, while the right-sided derivative $\frac{\partial^+ S}{y_j^{(i)}}$ is always zero. For $m = 2$ positive one-sided derivatives can be determined as the length of the segment of the attainment curve. Let $\mathbf{y}^{(i_L)}$ determine the neighbor of the weakly dominated point

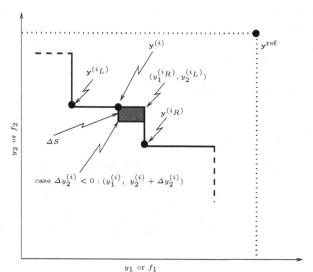

**Fig. 4.** Partial derivative for weakly dominated points in 2D. These points are dominated but not strictly dominated.

$\mathbf{y}^{(i)}$ on the upper left corner of the attainment curve, and $\mathbf{y}^{(i_R)}$ the neighbor on the lower right corner (see Fig. 4). If the point $\mathbf{y}^{(i)}$ lies on the segment $\mathbf{y}^{(i_L)}$ to $(y_1^{(i_R)}, y_2^{(i_L)})^\top$, then $\frac{\partial^+ \mathcal{S}}{\partial y_1^{(i)}} = 0$ and $\frac{\partial^- \mathcal{S}}{\partial y_2^{(i)}} = y_1^{(i_R)} - y_1^{(i)}$ (see also Fig. 4); else if the point lies on the segment $\mathbf{y}^{(i_R)}$ to $(y_1^{(i_R)}, y_2^{(i_L)})^\top$, then $\frac{\partial^- \mathcal{S}}{\partial y_1^{(i)}} = y_2^{(i_L)} - y_2^{(i)}$ and $\frac{\partial^+ \mathcal{S}}{\partial y_2^{(i)}} = 0$. The fact that $\mathcal{S}(\mathbf{p})$ is in general not continuously differentiable at weakly dominated points makes it problematic to work with gradient-based methods that make use of second order derivatives.

Weakly dominated points can also cause non-dominated points to have discontinuous local derivatives, which is comprehensible by arguments similar to the ones above. Besides degenerate points in the search space can cause discontinuous derivatives. These are, loosely defined, search points (or blocks) with the same image.

## 2.3   Empirical Determination of the Gradient

In practice the computation of the gradient can be approximated for example by using numerical differentiation. Since weakly non-dominated points of the population are not continuously differentiable, we need to take one-sided derivatives in both directions into account. For a small positive $\epsilon$ we compute them via:

$$\frac{\partial \mathcal{S}}{\partial p_i} \approx \frac{\mathcal{S}((p_1, \ldots, p_i \pm s\epsilon, \ldots, p_{\mu d})^\top) \pm \mathcal{S}((p_1, \ldots, p_i, \ldots, p_{\mu d})^\top)}{\epsilon} \qquad (11)$$

The algebraic signs we need to use depend on the gradients of the objective function. In case of continuously differentiable objective functions, it is numerically safer to compute the derivatives of the objective functions first, and then use the chain rule to compute the derivatives of the S-metric taking special care of weakly non-dominated points whenever they occur. Both the computation of Equation 11 and the computation of the gradients of all objective functions at all points (that can be used to compute the gradient via the chain rule) requires $\mu d$ evaluations of the objective function vectors.

## 3   Analytical Solution of S-Metric Maximization

We exemplarily verify the maximization of the S-metric with the gradient by an analytical calculation for a problem with a linear Pareto front $\{(y_1, y_2) \mid y_2 = 1 - y_1$ and $y_1 \in [0, 1]\}$ and a fixed number of points. Using analytical arguments and partial derivatives, the optimal positions of the points are calculated. Later we will use this problem and its solution for testing the local convergence behavior of the gradient-based method.

Due to the monotonicity of the S-metric the $\mu$ points of the approximation set that maximizes $S$ lie on the Pareto front. In order to consider the hypervolume of the approximation set we fix $(1, 1)$ as the reference point and we consider $\mu + 2$ points on this Pareto curve whose $y_1$-coordinates we denote by $u_i$, with $i = 0, \ldots, n + 1$. For any such collection of $n + 2$ points we always require $u_0 = 0$ and $u_{n+1} = 1$. We want to maximize the hypervolume with respect to $(1, 1)$. This is equivalent to minimizing the sum of the area of the triangles which are bounded by the Pareto curve and the sides of the rectangles shaping the attainment curve. Let $v_i$ denote the length of the interval between $u_i$ and $u_{i+1}$, then $\sum_{i=1}^{n+1} v_i^2$ is twice the area we want to minimize under the constraints $\sum_{i=1}^{n+1} v_i = 1$ and $\forall i : 0 \leq v_i$. This area is minimal in case the $n + 2$ points are uniformly distributed (with the understanding that two of the points are the end points). It is easy and worthwhile to prove this fact geometrically, yet we revert to an analytical verification as follows. Let $g := \sum_{i=1}^{n+1} v_i^2$. Incorporating the constraint $v_{n+1} = 1 - \sum_{i=1}^{n} v_i$ yields $g = \sum_{i=1}^{n} v_i^2 + (1 - \sum_{i=1}^{n} v_i)^2$. Computing the partial derivatives of $g$ results in $\frac{\partial g}{\partial v_j} = 2v_j - 2(1 - \sum_{i=1}^{n} v_i)$ where $j = 1, \ldots n$. Each of these partial derivatives has a value of zero at $v_1 = \frac{1}{n+1}, \ldots, v_n = \frac{1}{n+1}$ and at this point the minimum occurs. Translations back to the original problem result in $v_1 = \frac{1}{n+1}, \ldots, v_n = \frac{1}{n+1}$ and $v_{n+1} = \frac{1}{n+1}$. Hence, the points maximizing the S-metric are equidistant (with two occupying the end points).

Note that by approximating the Pareto front $\{(y_1, y_2) \mid y_i \in \mathbb{R}$ with $0 \leq y_1 \leq 1$ and $y_2 = 1 - y_2\}$ with a set consisting of $\mu$ points plus two extremal points $(0, 1), (1, 0)$ the maximal S-metric is $\frac{1}{2} \cdot \frac{\mu}{\mu+1}$. Moreover this maximum value can only be attained if the $\mu$ non-extremal points are equally spaced between the two extremal points.

With the generalized Schaffer problem Emmerich and Deutz [17] proposed a scalable-dimension problem that gives rise to the discussed linear Pareto front $\{(y, 1 - y) \mid y \in [0, 1]\}$ for $\alpha = 0.5$: $f_1(x) = \frac{1}{d^\alpha}(\sum_{i=1}^{d} x_i^2)^\alpha \rightarrow \min$ and

$f_2(x) = \frac{1}{d^\alpha}(\sum_{i=1}^d (1 - x_i)^2)^\alpha \to$ min for $x_i \in \mathbb{R}_+$, where $i = 1, ..., d$. In the following section, this problem and its solution set are consulted for a proof of concept result for the numerical optimization routines.

## 4    Gradient-Based Pareto Optimization

Due to the known problems with second-order gradient methods, which require twice continuous differentiability, a first-order gradient method, namely the steepest descent/ascent method with backtracking line search has been implemented [18]. The pseudo-code of our implementation is provided in Algorithm 1. The line-search algorithm has been kept simple to maintain transparency of the search process. It will however converge to a local maximizer relative to the line search direction. Note, that the line search may move to the same point in two subsequent iterations. In this case the evaluation of the objective function vectors of the population can be omitted. The convergence speed and accuracy of the line search can be controlled with the parameters $\tau$ and $\alpha_{min}$, respectively. We recommend a setting of $\tau = 0.1$, while the setting of $\alpha_{min}$ depends on the problem. Since the length of the gradient decreases when the algorithm converges to the optimum of a differentiable function, $\alpha_{min}$ does not have to be very low, because the length of the gradient influences the step-size as well.

---

**Algorithm 1. Gradient-ascent** S-metric maximization

---

1: **input variables**: initial population as $\mu d$ vector $\mathbf{p}$
2: **control variables**: accuracy of line search $\alpha_{min}$, step reduction rate $\tau \in (0, 1)$
3: $\alpha \leftarrow 1$ {Initialize step size $\alpha$}
4: $i \leftarrow 0; \mathbf{p}^{best} \leftarrow \mathbf{p}^0$
5: $\mathbf{d}^{(0)} \leftarrow \nabla \mathcal{S}(\mathbf{p}^{best})$ {Initialize search direction}
6: **while** $|\mathbf{d}^{(i)}| > \epsilon$ {Gradient larger than $\epsilon$} **do**
7:    $\alpha \leftarrow 1$
8:    **while** $\alpha > \alpha_{min}$ {Line search in gradient direction} **do**
9:       $\mathbf{p}^{new} \leftarrow \mathbf{p}^{best} + \alpha \mathbf{d}^{(i)}$ {Try positive direction}
10:      **if** $\mathcal{S}(\Psi(\mathbf{p}^{best})) \geq \mathcal{S}(\Psi(\mathbf{p}^{new}))$ **then**
11:         $\mathbf{p}^{new} \leftarrow \mathbf{p}^{best} - \alpha \mathbf{d}^{(i)}$ {Try negative direction}
12:         **if** $\mathcal{S}(\Psi(\mathbf{p}^{best})) \geq \mathcal{S}(\Psi(\mathbf{p}^{new}))$ {No success with both moves} **then**
13:            $\alpha \leftarrow \alpha \cdot \tau$ {Reduce step size $\alpha$}
14:            $\mathbf{p}^{new} \leftarrow \mathbf{p}^{best}$ {New current best point is old current best point}
15:         **end if**
16:      **end if**
17:      $\mathbf{p}^{best} \leftarrow \mathbf{p}^{new}$
18:   **end while**
19:   $\mathbf{d}^{(i+1)} \leftarrow \nabla \mathcal{S}(\mathbf{p}^{new}), i \leftarrow i + 1$ {Compute new gradient direction}
20: **end while**
21: **return** $\mathbf{p}^{best}$

---

## 5   SMS-EMOA-Gradient Hybrid

The gradient-descent method requires a good starting point in order to converge to the Pareto front. For this purpose an EMOA is applied which generates a good approximation of the Pareto front. We propose the SMS-EMOA because it has shown excellent results concerning the optimization of test functions and real-world problems (cf. [10,19,20]). The SMS-EMOA uses a steady-state selection scheme, i.e. in each generation one new solution is generated and one solution is discarded. A population of $\mu$ individuals is optimized without additional archives (which are often used in other EMOA). The S-metric is used within the selection operator to determine the subset of $\mu$ individuals with the highest S-metric value. Thereby, the individual with the least exclusive contribution of dominated hypervolume is discarded. As mentioned in Section 1, the maximization of the S-metric results in a well-distributed solution set with an emphasis of solutions in regions with fair trade-offs. The SMS-EMOA's final population functions as the starting point of the gradient strategy which does only a fine-tuning of the solutions. This sequential application of autonomous algorithms is called high-level relay hybridization according to the taxonomy introduced by Talbi [11]. The total number of function evaluations is partitioned among the algorithms.

*Experiment on the generalized Schaffer problem:* We conducted two experiments to analyze the limit behavior of the hybrid algorithm on the generalized Schaffer problem (Section 3) which reads $f_1(\mathbf{x}) = 1/d^\alpha(\sum_{i=1}^d x_i^2)^\alpha$, $f_2(\mathbf{x}) = 1/d^\alpha(\sum_{i=1}^d (1 - x_i)^2)^\alpha$, $x \in \mathcal{X} = [0,1]^d$, $\alpha \in \mathbb{R}^+$, and both objectives to be minimized. The first 1000 evaluations are always performed by SMS-EMOA. Figures 5 and 6 show a clipping of the subsequent behavior of typical runs, at which SMS-EMOA is always started using the same random seed.

In Fig. 5 the results pertaining to the generalized Schaffer problem with $d = 10$, $\alpha = \frac{1}{2}$ (hence the Pareto front is linear, cf. Section 3) of the following experiment are shown. The population size $\mu$ is 5, 10, or 15, and dimension $d$ is 10, 15, or 20. The purpose of this experiment was to study the convergence behavior of the gradient part of the algorithm. We see that the convergence (after a reasonable starting population has been found by the SMS part) is linear or almost linear. The former is especially true for small sizes of the approximation sets. The dimension of the search space has less effect on the speed of the methods. This can be explained by the relatively long time needed to perform line searches, as the dimension of the search space only influences the time needed for the gradient computation.

Fig. 6 shows the results for the generalized Schaffer problem with $\alpha = 1$, the dimension of the search space $d = 10$, and a population size (i.e., the size of the approximation set) of 10. The Pareto front is equal to $\{(y_1, y_2) \mid y_2 = 1 - 2\sqrt{y_1} + y_1 \text{ and } 0 \le y_1 \le 1\}$ and the maximally attainable S-metric is $1 - \frac{1}{6} \approx 0.833333$. The discontinuities in the progress correspond to the end of a line search, and a gap indicates that function evaluations are spend on the gradient calculation. The picture shows that once the gradient part of the hybrid method is supplied

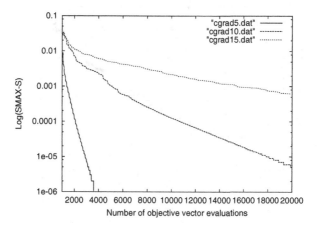

**Fig. 5.** The limit behavior of the gradient method starting from a population evolved over 1000 iterations with the SMS-EMOA for different problem dimensions $d$ and population sizes $\mu$. The logarithmic distance to the known optimum of the S-metric is plotted for different strategies.

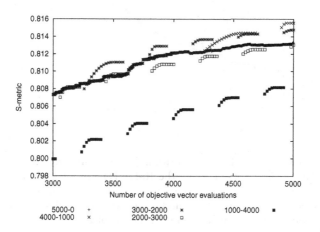

**Fig. 6.** The limit behavior of the gradient method starting from a population evolved over 1000 iterations by SMS-EMOA. The S-metric is plotted for different strategies, where the first number denotes the number of evaluations of the SMS part and the second of the gradient part.

with a reasonably good approximation set to the Pareto front the gradient part of the method outperforms the pure SMS-EMOA.

*Studies on the ZDT Test Suite:* Fig. 7 refers to the experiments run on the problem ZDT6 of the ZDT benchmark [5]. The size of the approximation set was chosen to be 20. Runs without penalty (Fig. 7, top) and with penalty (Fig. 7, bottom) on dominated points have been conducted. The total number of function evaluations

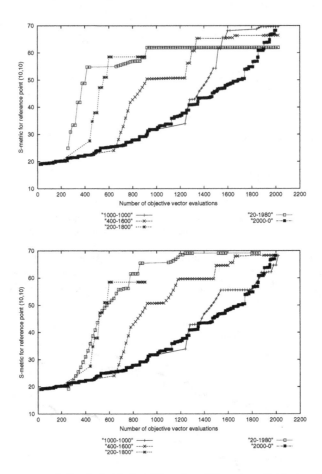

**Fig. 7.** Convergence of the hybrid algorithm for different switching points at which the gradient based solver takes over on the 10-D ZDT6 problem. In the upper (lower) figure the strategy is without (with) penalty. The numbers in the legend determine the number of function evaluation before and after switching. All strategies computed a total of 2000 evaluations and used a population size of 20.

in each run was 2000. Five different strategies were performed, listed with increasing number of function evaluations dedicated to the SMS part: 20, 200, 400, 1000, and 2000, respectively. The remainder of the 2000 function evaluations was used for the gradient part.

The two pictures reveal that it pays off to apply the gradient part of the algorithm as soon as a rough approximation set has been found. The speed-up occurs especially at the beginning and thus the hybrid approach is useful in case you would like to get very good results with few function evaluations. Secondly the picture also shows that giving a penalty to points in the population which are dominated gives far better approximation sets w.r.t. the S-metric.

**Table 1.** Runs with the relay hybrid obtained on the ZDT test suite. For each variant five runs have been performed. For ZDT4 the pure gradient approach failed to find a point dominating the reference point, thus the S-metric value remained zero.

| Problem | ST | MIN 1000 | AVG 1000 | MAX 1000 | MIN 1500 | AVG 1500 | MAX 1500 | MIN 2000 | AVG 2000 | MAX 2000 |
|---|---|---|---|---|---|---|---|---|---|---|
| ZDT1 | 1 | 21.876223 | 23.275247 | 24.199412 | 21.876223 | 23.850845 | 24.487564 | 21.876223 | 23.903422 | 24.505799 |
| ZDT1 | 2 | 21.118595 | 23.384878 | 24.342832 | 23.403862 | 23.974269 | 24.449935 | 23.604535 | 24.121280 | 24.457263 |
| ZDT1 | 3 | 17.202933 | 20.021265 | 21.845488 | 23.774588 | 24.101479 | 24.361921 | 24.175715 | 24.375030 | 24.482113 |
| ZDT1 | 4 | 17.202933 | 20.021265 | 21.845488 | 21.583049 | 22.643311 | 23.524577 | 24.113695 | 24.256499 | 24.403682 |
| ZDT1 | 5 | 17.202933 | 20.021265 | 21.845488 | 21.583049 | 22.643311 | 23.524577 | 23.060069 | 23.726008 | 24.365373 |
| ZDT2 | 1 | 19.162475 | 21.245294 | 24.064990 | 19.412864 | 22.237161 | 24.106690 | 19.412880 | 22.808071 | 24.127194 |
| ZDT2 | 2 | 18.126020 | 20.581421 | 23.101347 | 19.060637 | 21.727331 | 23.428600 | 19.695295 | 22.106164 | 23.962237 |
| ZDT2 | 3 | 14.661332 | 18.103744 | 19.873194 | 14.661332 | 20.404380 | 23.680670 | 14.661332 | 20.654252 | 23.686246 |
| ZDT2 | 4 | 14.661332 | 18.103744 | 19.873194 | 18.657467 | 19.972369 | 21.240527 | 19.999733 | 22.152473 | 23.640263 |
| ZDT2 | 5 | 14.661332 | 18.103744 | 19.873194 | 18.657467 | 19.972369 | 21.240527 | 19.995868 | 20.817935 | 22.205402 |
| ZDT3 | 1 | 22.399148 | 25.882488 | 27.149451 | 22.399148 | 26.109937 | 27.331745 | 22.399148 | 26.242184 | 27.373348 |
| ZDT3 | 2 | 24.461667 | 26.156686 | 27.223846 | 24.517026 | 26.290136 | 27.464974 | 24.535586 | 26.645181 | 27.488230 |
| ZDT3 | 3 | 19.142756 | 21.624740 | 23.348263 | 23.360338 | 25.773577 | 27.284952 | 23.797245 | 25.967965 | 27.398976 |
| ZDT3 | 4 | 19.142756 | 21.624740 | 23.348263 | 22.030572 | 24.145441 | 25.792775 | 23.555302 | 26.097489 | 27.32641 |
| ZDT3 | 5 | 19.142756 | 21.624740 | 23.348263 | 22.030572 | 24.145441 | 25.792775 | 24.719521 | 25.779436 | 27.260427 |
| ZDT4 | 1 | 0.000000 | 0.000000 | 0.000000 | 0.000000 | 0.000000 | 0.000000 | 0.000000 | 0.000000 | 0.000000 |
| ZDT4 | 2 | 1.991208 | 8.738555 | 12.296512 | 1.992431 | 9.560956 | 15.533157 | 1.993505 | 10.070364 | 17.979625 |
| ZDT4 | 3 | 6.070645 | 10.526529 | 14.042237 | 9.518876 | 12.649215 | 16.239738 | 10.526157 | 13.008094 | 16.239738 |
| ZDT4 | 4 | 6.070645 | 10.526529 | 14.042237 | 8.052115 | 11.965294 | 15.086376 | 8.052115 | 12.311928 | 16.310129 |
| ZDT4 | 5 | 6.070645 | 10.526529 | 14.042237 | 8.052115 | 11.965294 | 15.086376 | 8.587789 | 12.613113 | 16.121092 |
| ZDT6 | 1 | 57.826184 | 70.856967 | 78.100751 | 60.275019 | 72.771141 | 79.744496 | 60.364094 | 73.882299 | 83.297907 |
| ZDT6 | 2 | 36.830943 | 61.742804 | 72.663663 | 38.297947 | 68.691263 | 78.227012 | 51.547110 | 72.630236 | 79.137021 |
| ZDT6 | 3 | 38.012356 | 51.377244 | 63.308468 | 62.527358 | 68.822536 | 77.404330 | 71.326729 | 78.212762 | 85.345639 |
| ZDT6 | 4 | 38.012356 | 51.377244 | 63.308468 | 53.872935 | 75.230301 | 83.236934 | 75.959491 | 80.438820 | 85.029665 |
| ZDT6 | 5 | 38.012356 | 51.377244 | 63.308468 | 53.872935 | 75.230301 | 83.236934 | 81.736268 | 88.169766 | 91.648977 |

The finding of a reasonable approximation set to be used as a starting point for the gradient method is always done by the SMS. In the nearly pure gradient method also a very tiny fraction of the total number of functions evaluations is used by SMS-EMOA (20 evaluations). Clearly, the hybrid algorithm converges in each case to a population with maximum S-metric. Also the pure SMS method eventually catches up with the hybrid algorithm and converges to the maximum.

Table 1 shows the results of running the hybrid algorithm on the ZDT test suite (ZDT1 - ZDT4, and ZDT6). On each of the five problems the five different distributions of 2000 function evaluations among the hybrid parts are applied: (1) SMS: 20, gradient: 1980, (2) SMS: 500, gradient: 1500, (3) SMS: 1000 gradient: 1000, (4) SMS: 1500, gradient: 500, (5) SMS: 2000, gradient: 0. Each version of the hybrid algorithm is repeated five times with different random seeds. The reference point for each of the first four ZDTs was chosen as $(5, 5)$ and for ZDT6 it was $(10, 10)$. There are three checkpoints (at 1000, 1500, and 2000 evaluations) at which the minimal, average, and maximal S-metric are recorded (calculated concerning the five repetitions of a strategy). All strategies used the penalty function for dominated points. For ZDT1 and ZDT2 it is clear that the hybrid method is outperforming the pure SMS algorithm. In case of ZDT3 the pure gradient method is somewhat worse than the pure SMS on the other hand in case the first half of the function evaluations is spent on SMS (line 3 of ZDT3) the hybrid method outperforms the pure SMS again. A similar remark can be made about ZDT4 except that the pure gradient method in this case does not give good results due to reference point sensitivity. The reference point has been chosen too close to the Pareto front so that no point dominates it after a small number of function evaluations and the gradient strategy cannot work. The reference point sensitivity is not present in the SMS part of the algorithm as it only looks for relative increments of the hypervolume and (if $d = 2$) always selects extremal points directly. We see that when 500 or more evaluations are first spent on the SMS the hybrid is again competitive with the pure SMS. In case of ZDT6 which is multimodal the hybrid strategies do worse than the pure SMS. In all cases we see that the gradient method gives a speed-up especially in the beginning of the optimization.

## 6   Conclusions and Outlook

This paper introduces the gradient computation of the S-metric with respect to a population of points. Using the chain rule, the gradient of the S-metric can be computed from the gradients of the objective functions. It is important to distinguish between strictly dominated, weakly dominated, and non-dominated points. While for non-dominated sets differentiability is inherited from the objective functions, in the presence of weakly dominated points one-sided derivatives occur. For strictly dominated points sub-gradients with value zero occur. They make it impossible to improve these points by means of gradient methods. This problem can be partly circumvented by introducing a penalty approach.

However, the experiments in this paper show that it is advantageous to start the search with non-dominated sets close to the Pareto front, computed by an evolutionary algorithm, preferably one which maximizes the S-metric, too. Therefore, the proposed relay hybrid between the SMS-EMOA and a gradient method seems promising, though refined rules for phase switching still needs to be worked out. The study on the generalized Schaffer problem shows the potential of the new approach to find high precision approximations of finite populations maximizing the S-metric.

Future research should extend the empirical work on benchmarks and study problems of higher objective space dimension. Though some basic ideas of the gradient computation for more than two objectives using the chain rule have been sketched, details of the implementation need to be worked out.

## Acknowledgments

M. Emmerich acknowledges financial support by the Netherlands Organisation for Scientific Resarch (NWO). Nicola Beume is partly supported by the *Deutsche Forschungsgemeinschaft (DFG)* as part of the *Collaborative Research Center 'Computational Intelligence' (SFB 531)*.

## References

1. Zitzler, E., Thiele, L.: Multiobjective Optimization Using Evolutionary Algorithms—A Comparative Case Study. In: Eiben, A.E., Bäck, T., Schoenauer, M., Schwefel, H.-P. (eds.) PPSN V. LNCS, vol. 1498, pp. 292–301. Springer, Heidelberg (1998)
2. Zitzler, E., Brockhoff, D., Thiele, L.: The hypervolume indicator revisited: On the design of pareto-compliant indicators via weighted integration. In: Obayashi, S., Deb, K., Poloni, C., Hiroyasu, T., Murata, T. (eds.) EMO 2007. LNCS, vol. 4403, pp. 862–876. Springer, Heidelberg (2007)
3. Fleischer, M.: The measure of pareto optima. Applications to multi-objective metaheuristics. In: Fonseca, C.M., Fleming, P.J., Zitzler, E., Deb, K., Thiele, L. (eds.) EMO 2003. LNCS, vol. 2632, pp. 519–533. Springer, Heidelberg (2003)
4. Zitzler, E., Thiele, L., Laumanns, M., Fonseca, C.M., Grunert da Fonseca, V.: Performance assessment of multiobjective optimizers: An analysis and review. IEEE TEC 7(2), 117–132 (2003)
5. Deb, K.: Multi-Objective Optimization using Evolutionary Algorithms. Wiley, Chichester, UK (2001)
6. Coello Coello, C.A., Van Veldhuizen, D.A., Lamont, G.B.: Evolutionary Algorithms for Solving Multi-Objective Problems. Kluwer Academic Publishers, New York (2002)
7. Zitzler, E., Künzli, S.: Indicator-based selection in multiobjective search. In: Yao, X., Burke, E.K., Lozano, J.A., Smith, J., Merelo-Guervós, J.J., Bullinaria, J.A., Rowe, J.E., Tiňo, P., Kabán, A., Schwefel, H.-P. (eds.) PPSN VIII. LNCS, vol. 3242, pp. 832–842. Springer, Heidelberg (2004)
8. Huband, S., Hingston, P., While, L., Barone, L.: An evolution strategy with probabilistic mutation for multi-objective optimisation. In: CEC03, vol. 4, pp. 2284–2291. IEEE Computer Society Press, Los Alamitos (2003)

9. Knowles, J.: Local-Search and Hybrid Evolutionary Algorithms for Pareto Optimization. Phd thesis, Department of Computer Science, University of Reading, UK (2002)
10. Emmerich, M., Beume, N., Naujoks, B.: An EMO Algorithm Using the Hypervolume Measure as Selection Criterion. In: Coello Coello, C.A., Hernández Aguirre, A., Zitzler, E. (eds.) EMO 2005. LNCS, vol. 3410, pp. 62–76. Springer, Heidelberg (2005)
11. Talbi, E.G.: A Taxonomy of Hybrid Metaheuristics. Journal of Heuristics 8(5), 541–564 (2002)
12. Timmel, G.: Ein stochastisches Suchverfahren zur Bestimmung der Optimalen Kompromißlösungen bei statistischen polykriteriellen Optimierungsaufgaben. Journal TH Ilmenau 6, 139–148 (1980)
13. Fliege, J., Svaiter, B.F.: Steepest descent methods for multicriteria optimization. Mathematical Methods of Operations Research 51(3), 479–494 (2000)
14. Shukla, P., Deb, K., Tiwari, S.: Comparing Classical Generating Methods with an Evolutionary Multi-objective Optimization Method. In: Coello Coello, C.A., Hernández Aguirre, A., Zitzler, E. (eds.) EMO 2005. LNCS, vol. 3410, pp. 311–325. Springer, Heidelberg (2005)
15. Schütze, O., Dell'Aere, A., Dellnitz, M.: On continuation methods for the numerical treatment of multi-objective optimization problems. In: Branke, J., Deb, K., Miettinen, K., Steuer, R. (eds.) Practical Approaches to Multi-Objective Optimization. Dagstuhl Seminar Proceedings, IBFI, Schloss Dagstuhl, Germany, vol. 04461 (2005)
16. Bosman, P.A., de Jong, E.D.: Combining gradient techniques for numerical multiobjective evolutionary optimization. In: Keijzer, M., et al. (eds.) GECCO06, vol. 1, pp. 627–634. ACM Press, Seattle, USA (2006)
17. Emmerich, M., Deutz, A.: Test Problems based on Lamé Superspheres. In: Obayashi, S., Deb, K., Poloni, C., Hiroyasu, T., Murata, T. (eds.) EMO 2007. LNCS, vol. 4403, pp. 922–936. Springer, Heidelberg (2007)
18. Boyd, S., Vandenberghe, L.: Convex Optimization. Cambridge University Press, Cambridge, UK (2006)
19. Wagner, T., Beume, N., Naujoks, B.: Pareto-, Aggregation-, and Indicator-based Methods in Many-objective Optimization. In: Obayashi, S., Deb, K., Poloni, C., Hiroyasu, T., Murata, T. (eds.) EMO 2007. LNCS, vol. 4403, pp. 742–756. Springer, Heidelberg (2007)
20. Beume, N., Naujoks, B., Emmerich, M.: SMS-EMOA: Multiobjective Selection Based on Dominated Hypervolume. European Journal of Operational Research 181(3), 1653–1669 (2007)

# A Hybrid VNS for Connected Facility Location

Ivana Ljubić*

Department of Statistics and Decision Support Systems
University of Vienna
Austria
ivana.ljubic@univie.ac.at

**Abstract.** The connected facility location (ConFL) problem generalizes the facility location problem and the Steiner tree problem in graphs. Given a graph $G = (V, E)$, a set of customers $\mathcal{D} \subseteq V$, a set of potential facility locations $\mathcal{F} \subseteq V$ (including a root $r$), and a set of Steiner nodes in the graph $G = (V, E)$, a solution $(F, T)$ of ConFL represents a set of open facilities $F \subseteq \mathcal{F}$, such that each customer is assigned to an open facility and the open facilities are connected to the root via a Steiner Tree $T$. The total cost of the solution $(F, T)$ is the sum of the cost for opening the facilities, the cost of assigning customers to the open facilities and the cost of the Steiner tree that interconnects the facilities.

We show how to combine a variable neighborhood search method with a reactive tabu-search, in order to find sub-optimal solutions for large scale instances. We also propose a branch-and-cut approach for solving the ConFL to provable optimality. In our computational study, we test the quality of the proposed hybrid strategy by comparing its values to lower and upper bounds obtained within a branch-and-cut framework.

## 1 The Connected Facility Location Problem

Due to increasing customer demands regarding broadband connections, telecommunication companies search for solutions that "push" rapid and high-capacity fiber-optic networks closer to the subscribers, thus replacing the outdated copper twisted cable connections. The *Connected Facility Location Problem* (ConFL) models the next-generation of telecommunication networks: In the so-called *tree-star* networks, the core (fiber-optic) network represents a tree. This tree interconnects multiplexers that switch between fiber optic and copper connections. Each selected multiplexer is the center of the star-network of copper connections to its customers.

ConFL represents a generalization of two prominent combinatorial optimization problems: the *facility location problem* and the *Steiner tree problem* in graphs. More formally, ConFL is defined as follows: We are given an undirected graph $G = (V, E)$ with a set of *facilities* $\mathcal{F} \subseteq V$ and a set of *customer nodes* $\mathcal{D} \subseteq V$. We assign opening costs $f_i \geq 0$ to each facility $i \in \mathcal{F}$, edge costs $c_e \geq 0$

* Supported by the Hertha-Firnberg Fellowship of the Austrian Science Foundation (FWF).

T. Bartz-Beielstein et al. (Eds.): HM 2007, LNCS 4771, pp. 157–169, 2007.

to each edge $e \in E$, and demands $d_j$ to each customer $j \in \mathcal{D}$. We also assume that there is the set of Steiner nodes $\mathcal{S} = V \setminus (\mathcal{F} \cup \mathcal{D})$ and a root node $r \in \mathcal{F}$. Costs of assigning a customer $j \in \mathcal{D}$ to a facility $i \in \mathcal{F}$ are given as $a_{ij} \geq 0$. A solution $(F, T)$ of ConFL represents a set of open facilities $F \subseteq \mathcal{F}$, such that each customer $j \in \mathcal{D}$ is assigned to an open facility $i(j) \in F$ and the open facilities are connected to the root $r \in F$ by a Steiner Tree $T$. The total cost of the solution $(F, T)$ is the sum of the cost for opening the facilities, the cost of assigning customers to the open facilities and the cost of the Steiner tree that interconnects the facilities:

$$\sum_{i \in F} f_i + \sum_{j \in \mathcal{D}} d_j a_{i(j)j} + \sum_{e \in T} c_e.$$

In this constellation, some facility nodes may be used as pure Steiner nodes, in which case no opening costs for them will be paid.

As mentioned in [17], we can assume without loss of generality that the root $r$ represents an open facility and hence belongs to the Steiner tree. To solve the unrooted version of the problem, we simply need to run the algorithm for all facility nodes chosen as the root $r$. Without loss of generality, we can also assume that customer demands are all equal to one. Otherwise, we can set $a_{ij} \leftarrow d_j a_{ij}$ for all pairs $(i, j)$.

In Section 2, we propose a hybrid approach that combines Variable Neighborhood Search (VNS) with a reactive tabu search method. Section 3 describes a branch-and-cut (B&C) approach to solve the problem to provable optimality. In the last Section, our computational results show the comparison of the proposed hybrid approach with lower and upper bounds obtained within the B&C framework.

## 1.1   Related Work

ConFL has been introduced by Karger and Minkoff [6] who gave the first approximation algorithm of a constant factor. For the *metric* ConFL in which $c_e$ is a metric and $a_{ij} = 1/Mc_{ij}$, for some constant $M > 1$, Swamy and Kumar [17] used an integer linear programming (ILP) formulation to develop a primal-dual approximation algorithm of factor 8.55.

The *rent-or-buy* problem is a special case of ConFL in which there are no opening costs of the facilities and $\mathcal{F} = V$. A randomized 3.55-approximation algorithm for the metric rent-or-buy problem has been proposed by Gupta et al. [3]. The rent-or-buy problem has also been studied by Nuggehalli et al. [16] who gave a distributed greedy algorithm with approximation ratio 6, in the context of the design of ad hoc wireless networks. Their algorithm solves the rent-or-buy problem to optimality if the underlying graph has a tree topology.

The *Steiner tree-star* problem is another problem related to ConFL and to the node-weighted Steiner tree problem in graphs. The main difference to ConFL lies in the cost structure. To each non-customer node, we assign costs $f_i, i \in V \setminus \mathcal{D}$, assuming therefore that $\mathcal{F} = V \setminus \mathcal{D}$. If node $i$ belongs to the Steiner tree $T$, we pay for it no matter if any customer is assigned to it or not. Thus, the objective of the Steiner tree-star problem looks as follows:

$$\min \sum_{i \in T} f_i + \sum_{j \in \mathcal{D}} c_{i(j)j} + \sum_{e \in T} c_e,$$

where $c$ is the cost-function used both for assignments and for edge-costs.

Khuller and Zhu [7] gave a 5-approximation algorithm for solving the metric version of the problem. Lee et al. [11] proposed a branch-and-cut algorithm based on a separation of anti-cycle constraints. Their algorithm solved instances with up to 200 nodes to provable optimality. Xu et al. [18] developed a tabu search heuristic that incorporates long-term memory and probabilistic move selections. The authors considered insert-, delete-, and swap-moves, whereas swap-moves are used for diversification purposes. Computational results are given for instances of Lee et al. [11] and for additional sets of instances with up to 900 nodes. For the largest instances, the running time of the algorithm of more than 10 hours was reported.

Note that without the connectivity requirement (connecting the facilities by a Steiner tree), the ConFL reduces to the *uncapacitated facility location problem* (UFLP). On the other hand, if the set of facilities to be opened is known in advance, the problem is reduced to the Steiner tree problem in graphs (STP). Therefore, clearly, the problem is NP-hard.

## 2 A VNS with a Tabu Search

### 2.1 Basic VNS Model

According to Hoefer's computational study [5] on the uncapacitated facility location problem (UFLP), one of the successful metaheuristic approaches for solving the UFLP is a tabu-search approach given by Michel and van Hentenryck [15]. Recently, the authors obtained a very efficient strategy by simply extending a tabu-search approach with a variable neighborhood search. Our VNS framework follows these basic ideas given in [4]. Note however that, due to the nature of the problem, the way we calculate the objective value significantly differs from the one used to evaluate UFLP (see Section 2.2). Algorithm 1 shows our generic approach.

*Representation:* Assuming that for a fixed set of facilities, we can deterministically find a (sub)-optimal solution, we conduct our local search in the space of facility locations, thus changing configurations of vectors $y = (y_1, \ldots, y_{|\mathcal{F}|})$, where $y_i = 1$ if facility $i$ is open.

When it is clear from context, we will use $y$ to denote the subset of open facilities (i.e. those with $y_i = 1$).

*k-Neighborhood:* We define a $k$-neighborhood $N_k$ of a solution $\hat{y}$ by all solutions $y$ such that the Hamming distance $d(\hat{y}, y)$ between these two binary vectors is equal to $k$:

$$\mathcal{N}_k(\hat{y}) = \{ y \in \{0,1\}^{|\mathcal{F}|} \mid d(\hat{y}, y) = k \}.$$

---

**Data**    : Instance of the ConFL.
**Result** : A feasible suboptimal solution to ConFL.

$best = \hat{y} = Initialize()$;
$nIter = 0$;
**while** $nSame < LimitSame$ **and** $Time < TimeLimit$ **do**
    $y' = TabuSearch(\mathcal{N}_1(\hat{y}))$;
    $nIter + +$;
    **if** $Obj(y') > Obj(\hat{y})$ **then**
        $nSame + +$;
        $\hat{y} = Shake(y')$;
    **else**
        $\hat{y} = y'$;
        **if** $Obj(y') < Obj(best)$ **then**
            $nSame = 0$;
            Decrease the $tabuLength$;
            $best = y'$;
        **else**
            $nSame + +$;
            Increase the $tabuLength$;
        **end**
    **end**
**end**
**return** $best$;

---

**Algorithm 1.** VNS algorithm

*Reactive Tabu Search:* The status of a facility $i \in \mathcal{F}$ is given by $y_i$. A basic *move* is the change of the status of a facility, i.e. $y_i \leftarrow 1 - y_i$. The tabu list consists of the set of facilities that cannot be flipped. A solution $\hat{y}$ is locally improved using the *best improvement* strategy with respect to its 1-neighborhood. Thus, all possible flips of single positions that are not in the tabu list are considered, and the best one is taken. If there is more than one best flip, we randomly select one.

In order to forbid the reversal of recent search steps, we use a self-learning mechanism that adapts the length of the tabu list during the search. We simplify the ideas of the *reactive tabu search*, which was originally proposed by Battiti and Tecchiolli [1]: the list size is increased whenever no improvement upon the best found solution is made. Whenever a new best solution is detected, the list size is decreased.

We implement a dynamic tabu list in the following way: to each facility $i \in \mathcal{F}$, we associate a counter $tabuList(i)$. When a facility is inserted into the tabu list, we set $tabuList(i) \leftarrow nIter + tabuLength$, which forbids flipping the facility $i$ for the next $tabuLength$ iterations, whereby $nIter$ denotes the current iteration. The value of $tabuLength$ is adjusted automatically: if tabu search improves the value of $\hat{y}$, but it is still worse than the best obtained value, we increase the length of the tabu list by one. Otherwise, the length of the tabu list will be decreased by one. We use standard settings for minimal and maximal values of *tabuLength*, and set them to 2 and 10, respectively.

*Shaking the Neighborhood:* This diversification mechanism enables escaping from local optima found within the tabu search procedure. If the last tabu search iteration did not improve upon the last selected value $\hat{y}$, we randomly select $k$ ($k \geq 2$) positions and flip them. The value $k$ increases until it reaches a pre-specified maximum neighborhood size (50 in the default implementation), after which it starts from 2 again.

Using this technique, the diversification degree will be automatically adjusted. Increasing the size of neighborhoods systematically also assures that significant diversifications are avoided during early phases of the search.

*Hash-Tables:* Since the evaluation of solutions is computationally expensive (see next Subsection), we maintain hash-tables for all vectors $y$ for which the objective value for STP, assignment or total ConFL value is already known (see also [10], where a similar idea has been used). This strategy ensures that the objective value of the same vector will not be calculated more than once within the whole procedure, even if we return back to the same solution.

*Termination Criteria:* The algorithm terminates if the best found solution was not improved within the last *LimitSame* iterations, or a pre-specified *TimeLimit* is exceeded. In our default implementation, we set *LimitSame* to 50, and *TimeLimit* to 1000 seconds.

## 2.2  Evaluation of the Objective Function

Algorithm 2 shows the main steps of calculating the objective function for a specified vector $\hat{y}$. We use the following notation:

- vectors $x$ ( $= x^P$ or $x^A$) refer to the assignment values, i.e. $x_{ij} = 1$ if customer $j$ is assigned to facility $i$ and $x_{ij} = 0$, otherwise;
- $T^P$ and $T^{MST}$ denote the sets of nodes and edges building a Steiner tree that connects the chosen set of facilities ($y^P$ and $y^{MST}$, resp.).

Given $\hat{y}$, we first check if this configuration has been already calculated before. If so, we get the corresponding tree-, assignment-, and facility values from the hash-table *Hash*. Otherwise, we run a three-step procedure:

**Step 1:** $(T^{MST}, y^{MST}) = MSTHeuristic(\hat{y})$: We consider the graph $G' = (V', E')$ – a subgraph of $G$ induced by the set of facilities and Steiner nodes $V' = \mathcal{F} \cup \mathcal{S}$ with the edge costs $c$. For $G'$, we generate the so-called *distance network*[1] - a complete graph whose nodes correspond

---

[1] Calculation of the distance network is done only once in the beginning of the VNS algorithm.

---

**Data**     : Vector $\hat{y}$: a facility $i$ is selected if $\hat{y}_i = 1$.
**Result** : Locally improved vector $y^P$ and its objective function value.

**if** $Hash(\hat{y})$ *defined* **then**
    $(x^P, T^P, y^P) = Hash(\hat{y})$;
**else**
    $(T^{MST}, y^{MST}) = MSTHeuristic(\hat{y})$;
    $(x^A, y^A) = Assign(y^{MST})$;
    $(x^P, T^P, y^P) = Peeling(T^{MST}, x^A, y^A)$;
    Insert $(x^P, T^P, y^P)$ into $Hash$;
**end**
**return** $\sum_{e \in T_P} c_e + \sum_{i \in \mathcal{F}} f_i y_i^P + \sum_{i \in \mathcal{F}} \sum_{j \in \mathcal{D}} a_{ij} x_{ij}^P$;

---

**Algorithm 2.** Calculating the objective function

to facilities $i \in \mathcal{F}$, and whose edge-lengths $l(i,j)$ are defined as shortest paths in $G'$, for all $i, j \in \mathcal{F}$.

We use the minimum spanning tree (MST) heuristic [14] to find a spanning tree $T^{MST}$ that connects all selected facilities ($\hat{y}_i = 1$).

1. Let $G''$ be the subgraph of $G'$ induced by $\hat{y}$.
2. Calculate the minimum spanning tree $MST_G''$ of the distance sub-network $G''$.
3. Include in $T^{MST}$ all intermediate edges and nodes of $G$ contained in selected shortest-path edges from $MST_G''$.
4. Update the set of selected facilities: set $y_i^{MST} = 1$ for all nodes $i \in T^{MST} \cap \mathcal{F}$.

**Step 2:** $(x^A, y^A) = Assign(y^{MST})$: For each customer $j \in \mathcal{D}$, we find the cheapest possible assignment to a facility from $y^{MST}$. The values are stored in vector $x^A$. Since not every facility from $y^{MST}$ necessarily serves a customer, we denote with $y^A$ the subset of those that really need to be opened.

This operation is calculated from scratch – although the differences between two neighboring vectors $\hat{y}_1$ and $\hat{y}_2$ are in general very small, the corresponding $y_1^{MST}$ and $y_2^{MST}$ solutions may be significantly different. Thus, the total computational complexity for finding the cheapest assignment in the worst case is $O(|\mathcal{F}||\mathcal{D}|)$.

**Step 3:** $(x^P, T^P, y^P) = Peeling(T^{MST}, x^A, y^A)$: We finally want to get rid of some of those facilities that are still part of the Steiner tree, but that are not used at all. We do this by applying the so-called *peeling procedure*. Our peeling heuristic tries to recursively remove all redundant leaf nodes (including corresponding tree-paths) from the tree-solution $T^{MST}$. Let $k$ denote a leaf node, and let $P_k$ be a path that connects $k$ to the next open facility from $y^A$, or to the next branch, towards the root $r$.

1. If the leaf node is not an (open) facility, i.e. if $k \notin y^A$, we simply delete $P_k$.

2. Otherwise, we try to to re-assign customers (originally assigned to $k$) to already open facilities (if possible). If such obtained solution is better, we delete $P_k$ and continue processing other leaves.

The main steps of this procedure are given in Algorithm 3.

If the set of facilities is sorted for each customer in increasing order with respect to its assignment costs[2], this procedure can be implemented very efficiently. Indeed, in order to find an open facility (from $y^P$) nearest to $j$ and different from $k$ (denoted by $i^k(j)$), we only need to proceed this ordered list starting from $k$ until we encounter a facility from $y^P$.

The algorithm stops when only one node is left, or when all the leaves from $T^P$ have been proceeded. Thus, the worst-case running time of the whole peeling method is $O(|\mathcal{F}||\mathcal{D}|)$.

---

**Data** : Set $y^A$, assignment $x^A$ and a Steiner tree $T^{MST}$.
**Result** : Locally improved vector $y^P$ and its objective function value.

$T^P = T^{MST}, y^P = y^{MST}, x^P = x^A$;
**for** *all leaves $k$ in $T^P$* **do**
    Determine path $P_k$ and its costs $c(P_k) = \sum_{e \in P_k} c_e$;
    **if** $k \notin y^P$ **then**
        $T^P = T^P - P_k$;
    **else**
        $\mathcal{D}_k = \{j \mid j \in \mathcal{D}, x^P_{kj} = 1\}$;
        $i^k(j) = \arg\min\{a_{ij} \mid i \in y^P, i \neq k\}, \forall j \in \mathcal{D}_k$;
        **if** $\sum_{j \in \mathcal{D}_k} a_{i^k(j)j} < f_k + c(P_k) + \sum_{j \in \mathcal{D}_k} a_{kj}$ **then**
            $y^P_k = 0$;
            $T^P = T^P - P_k$;
            $x^P_{kj} = 0, x^P_{i(k)j} = 1, \forall j \in \mathcal{D}_k$;
        **end**
    **end**
**end**

**Algorithm 3.** Peeling procedure

## 3   Branch-and-Cut for ConFL

We propose to calculate lower bounds and provably optimal solutions to ConFL using the integer linear programming (ILP) model given below. For solving the

---

[2] Sorting of these lists is done once, in the initialization phase of VNS algorithm.

linear programming relaxations and for a generic implementation of the branch-and-cut approach, we used the commercial packages ILOG CPLEX (version 10.0) and ILOG Concert Technology (version 2.2).

We solve the ConFL to optimality on a directed graph $G_A = (V, A)$ obtained from the original one $G = (V, E)$ by simply replacing each edge $e \in E$ by two directed arcs of the same cost:

$$A = \{(k, l) | \{k, l\} \in E \wedge l \neq r\}$$

$$c_{kl} = c(\{k, l\}), \quad \forall (k, l) \in A$$

The assignment costs $(a_{ij})$ remain unchanged.

The problem of finding a rooted Steiner tree on a directed graph is known as the *Steiner arborescence problem*: given $G_A$, a root $r$ and the set of terminals $F \subset V$, find a subset of arcs $R \subset A$ such that there is a directed path from $r$ to each $i \in F$, and that $\sum_{(k,l) \in R} c_{kl}$ is minimized.

To model the problem, we use the following binary vectors: $y_i$ indicates whether a facility $i$ is open, $x_{ij}$ indicates whether customer $j$ is assigned to facility $i$ and $z_{kl}$ indicates whether the arc $(k, l)$ is a part of the directed Steiner tree rooted at $r$.

$$(ConFL) \quad \min \quad \sum_{i \in \mathcal{F}} f_i y_i + \sum_{i \in \mathcal{F}} \sum_{j \in \mathcal{D}} a_{ij} x_{ij} + \sum_{(k,l) \in A} c_{kl} z_{kl} \qquad (1)$$

$$\sum_{i \in \mathcal{F}} x_{ij} \geq 1, \qquad \forall j \in \mathcal{D} \qquad (2)$$

$$x_{ij} \leq y_i, \qquad \forall i \in \mathcal{F}\, \forall j \in \mathcal{D} \qquad (3)$$

$$\sum_{(k,l) \in \delta^-(S)} z_{kl} \geq y_i, \forall S \subseteq V \setminus \{r\}, i \in S \cap \mathcal{F} \neq \emptyset \qquad (4)$$

$$y_r = 1 \qquad (5)$$

$$0 \leq x_{ij}, z_{kl}, y_i \leq 1 \quad \forall i,\, \forall j,\, \forall (k, l) \in A \qquad (6)$$

$$x_{ij}, z_{kl}, y_i \in \{0, 1\} \quad \forall i,\, \forall j,\, \forall (k, l) \in A \qquad (7)$$

Here, with $\delta^-(S)$ we denote the set of ingoing edges of $S$, i.e., $\delta^-(S) = \{(k, l) \in A \mid k \notin S, l \in S\}$.

The *assignment constraints* (2) ensure that each customer is assigned to exactly one facility. The *capacity constraints* (3) ensure that customers can only be assigned to open facilities. The *connectivity constraints* (4) guarantee that there is a directed path between every open facility and the root $r$, i.e. they ensure that open facilities are connected to the root and to each other. With constraint (5) we fix the root node $r$. Constraints (4) and (5) ensure existence of the Steiner arborescence, whereas constraints (2) and (3) ensure a feasible assignment.

*Initialization:* We initialize the LP with relaxed integer requirements (6), with assignment- and capacity-inequalities (2)-(3), with indegree inequalities:

$$\sum_{(k,l) \in A} z_{kl} = y_l, \forall l \in \mathcal{F}$$

and with the *subtour elimination constraints* of size two:

$$z_{kl} + z_{lk} \leq y_l, \quad \forall l \in \mathcal{F}.$$

Additionally, we add flow-balance constraints ([8]) that ensure that the in-degree of each Steiner node is less or equal than its out-degree:

$$\sum_{(k,l)\in A} z_{kl} \leq \sum_{(l,k)\in A} z_{lk}, \quad \forall l \notin \mathcal{F}.$$

*Separation of Cut Inequalities:* In each node of the branch-and-bound tree we separate the cut-inequalities given by (4). For a given LP-solution $\hat{z}$, we construct a support graph $G_{\hat{z}} = (V, A, z)$ with arc-weights $\hat{z} : A \mapsto [0, 1]$. Then we calculate the minimum cost flow from the root $r$ to each potential facility node $i \in \mathcal{F}$ such that $y_i > 0$. If this min-cost flow value is less than $y_i$, we have a violated inequality, induced by the corresponding min-cut in the graph $G_{\hat{z}}$, and we insert it into the LP.

To improve computational efficiency, we search for *nested, back* and *minimum-cardinality cuts* and insert at most 100 violated inequalities in each separation phase. For more details, see our implementation of the B&C algorithm for the prize-collecting Steiner tree problem, where the same separation procedure has been used [12,13].

*Branching:* Branching on single arc variables produces a huge disbalance in the branch-and-bound tree. Whereas discarding an edge from the solution (setting $z_{kl}$ to zero) doesn't bring much, setting the node variable to one, significantly reduces the size of the search subspace. Therefore we set the highest branching priority to potential facility nodes $i \in \mathcal{F}$.

## 4   Computational Results

We consider three classes of benchmark instances, obtained by merging data from three public sources. In general, we combine an UFLP instance with an STP instance, to generate ConFL input graphs in the following way: first $|\mathcal{F}|$ nodes of the STP instance are selected as potential facility locations, and the node with index 1 is selected as the root. The number of facilities, the number of customers, opening costs and assignment costs are provided in UFLP files. STP files provide edge-costs and additional Steiner nodes.

  - We consider two sets of non-trivial UFLP instances from UflLib[3]:
    • MP-{1,2} and MQ-{1,2} instances have been proposed by Kratica et al. [10]. They are designed to be similar to UFLP real-world problems and have a large number of near-optimal solutions. There are 6 classes of problems, and for each problem $|\mathcal{F}| = |\mathcal{D}|$. We took 2 representatives of the 2 classes MP and MQ of sizes $200 \times 200$ and $300 \times 300$, respectively.

---

[3] http://www.mpi-inf.mpg.de/departments/d1/projects/benchmarks/UflLib/

- The GS-{250,500}-{1,2} benchmark instances were initially proposed by Koerkel [9] (see also Ghosh [2]). Here we chose two representatives of the 250 × 250 and 500 × 500 classes, respectively. Connection costs are drawn uniformly at random from [1000, 2000], while opening costs are drawn uniformly at random from [100, 200].

- STP instances:
  - Instances {C,D}5, {C,D}10, {C,D}15, {C,D}20 were chosen randomly from the OR-library[4] as representatives of medium size instances for the STP.

All experiments were performed on a Pentium D, 3.0 GHz machine with 2GB RAM. The first table shows the number of facility nodes ($|\mathcal{F}|$), the number of customers ($|\mathcal{D}|$), and the number of Steiner nodes ($|\mathcal{S}|$); because sets are disjoint, $\mathcal{S} = V \setminus (\mathcal{D} \cup \mathcal{F})$. Furthermore, lower bounds (LB) and upper bounds (UB) obtained after running the B&C algorithm for one hour are provided.

The number of nodes in the branch-and-bound tree and the running time of the exact method indicate that the instances with no more than 300 customer- and facility nodes are not trivial, but also not too difficult for the selected method.

For 15 out of 48 benchmark instances – 6 from the first and 9 from the second group – our B&C algorithm finds an optimal solution in less than one hour. Note that for the rest of the instances, we provide upper bounds found by local improvement methods already incorporated in the CPLEX solver 10.0, without using any additional primal heuristics.

The second table shows average and best values (out of 10 runs) obtained from running the VNS strategy with time limit of 1000 seconds. Initial solutions are obtained by randomly selecting 5% of potential facilities.

We provide the best found value of the VNS approach, as well as the best- and average-gaps out of 10 runs ($gap_{best}$ and $gap_{avg}$, resp.). Standard deviation of the gap is given in column $gap_{stddev}$. The average number of iterations, and the average running time (in seconds) needed to detect the best solution of each run are given in the last two columns. Note that the gap values are always calculated with respect to the lower bound given in the first column.

The obtained results clearly indicate that the B&C algorithm is not able to handle instances with a large number of customer- or facility nodes within a reasonable amount of time. Already for instances with $\mathcal{F} = \mathcal{D} = 500$, the algorithm is not able to close the optimality gap. The main difficulty for B&C (and exact methods in general) comes from the assignment and capacity constraints. On the other side, for the same set of instances, our VNS approach finds solutions which are within 1% of the lower bound.

For the instances of the first two groups, the algorithm does not always reach the optimal solution, but the average gaps and their standard deviation indicate a stable performance and the robustness of the approach.

The incorporation of the VNS method as a primal heuristic within the B&C framework seems a promising direction for further research. The synergy effect

---

[4] http://people.brunel.ac.uk/~mastjjb/jeb/orlib/steininfo.html

**Table 1.** Lower and upper bounds of selected benchmark instances obtained by running B&C algorithm with time limit of one hour

| Instances | | Properties | | | | | B&C Results | | | |
|---|---|---|---|---|---|---|---|---|---|---|
| UFLP | STP | $|\mathcal{D}|$ | $|\mathcal{F}|$ | $|\mathcal{S} \cup \mathcal{F}|$ | $|E|$ | $|\mathcal{S}|$ | LB | UB | Time | B&Bnodes |
| mp1 | c5 | 200 | 200 | 500 | 625 | 300 | 2868.6 | 2889.6 | 3602.1 | 3458 |
| mp2 | c5 | 200 | 200 | 500 | 625 | 300 | 2869.5 | 2869.5 | 818.4 | 76 |
| mp1 | c10 | 200 | 200 | 500 | 1000 | 300 | 2672.1 | 2693.9 | 3602.2 | 2496 |
| mp2 | c10 | 200 | 200 | 500 | 1000 | 300 | 2663.5 | 2663.5 | 185.8 | 62 |
| mp1 | c15 | 200 | 200 | 500 | 2500 | 300 | 2636.7 | 2636.7 | 332.4 | 82 |
| mp2 | c15 | 200 | 200 | 500 | 2500 | 300 | 2646.5 | 2646.5 | 226.3 | 12 |
| mp1 | c20 | 200 | 200 | 500 | 12500 | 300 | 2606.7 | 2619.7 | 3603.4 | 1059 |
| mp2 | c20 | 200 | 200 | 500 | 12500 | 300 | 2627.5 | 2627.5 | 172.6 | 102 |
| mq1 | c5 | 300 | 300 | 500 | 625 | 200 | 4357.8 | 4357.8 | 2740.8 | 50 |
| mq2 | c5 | 300 | 300 | 500 | 625 | 200 | 3770.8 | 4185.6 | 3602.8 | 61 |
| mq1 | c10 | 300 | 300 | 500 | 1000 | 200 | 3878.4 | 3971.2 | 3602.9 | 746 |
| mq2 | c10 | 300 | 300 | 500 | 1000 | 200 | 3697.6 | 3778.8 | 3602.6 | 1025 |
| mq1 | c15 | 300 | 300 | 500 | 2500 | 200 | 3778.9 | 3868.8 | 3603.1 | 500 |
| mq2 | c15 | 300 | 300 | 500 | 2500 | 200 | 3612.7 | 3692.6 | 3603.0 | 918 |
| mq1 | c20 | 300 | 300 | 500 | 12500 | 200 | 3738.4 | 3830.5 | 3603.9 | 277 |
| mq2 | c20 | 300 | 300 | 500 | 12500 | 200 | 3641.7 | 3689.5 | 3603.9 | 268 |
| mp1 | d5 | 200 | 200 | 1000 | 1250 | 800 | 2704.9 | 2704.9 | 445.6 | 208 |
| mp2 | d5 | 200 | 200 | 1000 | 1250 | 800 | 2759.6 | 2759.6 | 3394.3 | 1869 |
| mp1 | d10 | 200 | 200 | 1000 | 2000 | 800 | 2667.9 | 2678.9 | 3607.7 | 2315 |
| mp2 | d10 | 200 | 200 | 1000 | 2000 | 800 | 2688.5 | 2688.5 | 1777.7 | 1379 |
| mp1 | d15 | 200 | 200 | 1000 | 5000 | 800 | 2639.7 | 2639.7 | 1089.7 | 269 |
| mp2 | d15 | 200 | 200 | 1000 | 5000 | 800 | 2648.5 | 2648.5 | 174.4 | 14 |
| mp1 | d20 | 200 | 200 | 1000 | 25000 | 800 | 2620.7 | 2620.7 | 971.8 | 410 |
| mp2 | d20 | 200 | 200 | 1000 | 25000 | 800 | 2628.5 | 2628.5 | 2919.2 | 1011 |
| mq1 | d5 | 300 | 300 | 1000 | 1250 | 700 | 3919.1 | 3919.1 | 3230.4 | 417 |
| mq2 | d5 | 300 | 300 | 1000 | 1250 | 700 | 3721.5 | 3835.7 | 3609.1 | 478 |
| mq1 | d10 | 300 | 300 | 1000 | 2000 | 700 | 3765.7 | 3945.1 | 3609.2 | 220 |
| mq2 | d10 | 300 | 300 | 1000 | 2000 | 700 | 3627.3 | 3749.3 | 3608.7 | 293 |
| mq1 | d15 | 300 | 300 | 1000 | 5000 | 700 | 3844.5 | 3844.5 | 3574.1 | 768 |
| mq2 | d15 | 300 | 300 | 1000 | 5000 | 700 | 3609.0 | 3710.7 | 3609.3 | 387 |
| mq1 | d20 | 300 | 300 | 1000 | 25000 | 700 | 3751.6 | 3846.8 | 3614.4 | 119 |
| mq2 | d20 | 300 | 300 | 1000 | 25000 | 700 | 3597.0 | 3752.0 | 3614.4 | 63 |
| gs250a-1 | c5 | 250 | 250 | 500 | 625 | 250 | 258300.0 | 259067.0 | 3602.8 | 370 |
| gs250a-2 | c5 | 250 | 250 | 500 | 625 | 250 | 257920.0 | 258714.0 | 3602.8 | 141 |
| gs250a-1 | c10 | 250 | 250 | 500 | 1000 | 250 | 258062.0 | 258855.0 | 3604.1 | 80 |
| gs250a-2 | c10 | 250 | 250 | 500 | 1000 | 250 | 257685.0 | 258450.0 | 3606.8 | 99 |
| gs250a-1 | c15 | 250 | 250 | 500 | 2500 | 250 | 257843.0 | 258307.0 | 3603.4 | 205 |
| gs250a-2 | c15 | 250 | 250 | 500 | 2500 | 250 | 257524.0 | 257922.0 | 3604.3 | 106 |
| gs250a-1 | c20 | 250 | 250 | 500 | 12500 | 250 | 257815.0 | 258494.0 | 3603.8 | 51 |
| gs250a-2 | c20 | 250 | 250 | 500 | 12500 | 250 | 257452.0 | 258253.0 | 3604.1 | 22 |
| gs500a-1 | c5 | 500 | 500 | 500 | 625 | 0 | 511499.0 | 756415.0 | 3607.0 | 0 |
| gs500a-2 | c5 | 500 | 500 | 500 | 625 | 0 | 511485.0 | 756031.0 | 3614.2 | 0 |
| gs500a-1 | c10 | 500 | 500 | 500 | 1000 | 0 | 511056.0 | 666907.0 | 3607.8 | 0 |
| gs500a-2 | c10 | 500 | 500 | 500 | 1000 | 0 | 511018.0 | 756031.0 | 3614.1 | 0 |
| gs500a-1 | c15 | 500 | 500 | 500 | 2500 | 0 | 510733.0 | 665029.0 | 3606.6 | 0 |
| gs500a-2 | c15 | 500 | 500 | 500 | 2500 | 0 | 510718.0 | 671085.0 | 3606.8 | 0 |
| gs500a-1 | c20 | 500 | 500 | 500 | 12500 | 0 | 510584.0 | 668346.0 | 3611.5 | 0 |
| gs500a-2 | c20 | 500 | 500 | 500 | 12500 | 0 | 510559.0 | 600688.0 | 3608.7 | 0 |

**Table 2.** Comparison of the VNS with lower and upper bounds obtained by B&C

| Instances | | B&C | | VNS | | | | | |
|---|---|---|---|---|---|---|---|---|---|
| UFLP | STP | LB | UB-gap | best | $gap_{best}$ | $gap_{avg}$ | $gap_{stddev}$ | Iter | Time |
| mp1 | c5 | 2868.6 | 0.7 | 2923.7 | 1.9 | 2.1 | 0.1 | 22.1 | 98.6 |
| mp2 | c5 | 2869.5 | **0.0** | 2880.4 | 0.4 | 0.4 | 0.0 | 66.8 | 363.1 |
| mp1 | c10 | 2672.1 | 0.8 | 2799.6 | 4.8 | 5.2 | 0.4 | 76.0 | 389.0 |
| mp2 | c10 | 2663.5 | **0.0** | 2743.3 | 3.0 | 3.0 | 0.0 | 40.0 | 187.2 |
| mp1 | c15 | 2636.7 | **0.0** | 2731.2 | 3.6 | 4.0 | 1.2 | 78.3 | 406.4 |
| mp2 | c15 | 2646.5 | **0.0** | 2801.2 | 5.8 | 6.7 | 1.4 | 35.7 | 186.1 |
| mp1 | c20 | 2606.7 | 0.5 | 2665.6 | 2.3 | 2.5 | 0.6 | 50.5 | 300.3 |
| mp2 | c20 | 2627.5 | **0.0** | 2700.6 | 2.8 | 3.3 | 0.6 | 52.6 | 305.2 |
| mq1 | c5 | 4357.8 | **0.0** | 4474.6 | 2.7 | 2.7 | 0.0 | 21.2 | 177.7 |
| mq2 | c5 | 3770.8 | 11.0 | 4185.6 | 11.0 | 11.9 | 1.4 | 40.0 | 353.0 |
| mq1 | c10 | 3878.4 | 2.4 | 4037.8 | 4.1 | 7.4 | 4.9 | 40.2 | 365.0 |
| mq2 | c10 | 3697.6 | 2.2 | 3934.9 | 6.4 | 7.2 | 1.7 | 36.7 | 340.1 |
| mq1 | c15 | 3778.9 | 2.4 | 4058.4 | 7.4 | 8.8 | 1.7 | 48.5 | 528.9 |
| mq2 | c15 | 3612.7 | 2.2 | 3743.0 | 3.6 | 3.6 | 0.1 | 25.9 | 250.5 |
| mq1 | c20 | 3738.4 | 2.5 | 3996.7 | 6.9 | 8.2 | 2.0 | 38.1 | 401.2 |
| mq2 | c20 | 3641.7 | 1.3 | 3782.8 | 3.9 | 5.1 | 0.6 | 35.8 | 375.5 |
| mp1 | d5 | 2704.9 | **0.0** | 2715.5 | 0.4 | 2.2 | 1.0 | 23.5 | 402.9 |
| mp2 | d5 | 2759.6 | **0.0** | 2766.2 | 0.2 | 1.4 | 1.2 | 31.7 | 482.3 |
| mp1 | d10 | 2667.9 | 0.4 | 2728.8 | 2.3 | 3.4 | 1.3 | 23.1 | 366.8 |
| mp2 | d10 | 2688.5 | **0.0** | 2691.5 | 0.1 | 0.6 | 0.8 | 13.6 | 365.0 |
| mp1 | d15 | 2639.7 | **0.0** | 2694.6 | 2.1 | 2.3 | 0.7 | 23.0 | 328.5 |
| mp2 | d15 | 2648.5 | **0.0** | 2705.6 | 2.2 | 2.8 | 1.3 | 20.4 | 379.0 |
| mp1 | d20 | 2620.7 | **0.0** | 2634.9 | 0.5 | 0.9 | 0.8 | 24.0 | 453.4 |
| mp2 | d20 | 2628.5 | **0.0** | 2629.5 | 0.0 | 0.0 | 0.0 | 15.4 | 321.9 |
| mq1 | d5 | 3919.1 | **0.0** | 4012.3 | 2.4 | 6.6 | 4.3 | 22.1 | 508.1 |
| mq2 | d5 | 3721.5 | 3.1 | 4007.3 | 7.7 | 11.8 | 2.5 | 21.4 | 460.7 |
| mq1 | d10 | 3765.7 | 4.8 | 3975.0 | 5.6 | 7.0 | 1.5 | 15.3 | 511.1 |
| mq2 | d10 | 3627.3 | 3.4 | 3796.4 | 4.7 | 7.0 | 1.4 | 18.0 | 593.8 |
| mq1 | d15 | 3844.5 | **0.0** | 3857.6 | 0.3 | 7.0 | 8.6 | 24.5 | 652.8 |
| mq2 | d15 | 3609.0 | 2.8 | 3798.9 | 5.3 | 8.9 | 3.8 | 27.9 | 627.0 |
| mq1 | d20 | 3751.6 | 2.5 | 3984.6 | 6.2 | 9.1 | 2.1 | 17.9 | 490.4 |
| mq2 | d20 | 3597.0 | 4.3 | 3749.0 | 4.2 | 5.4 | 1.5 | 17.6 | 495.5 |
| gs250a-1 | c5 | 258300.0 | 0.3 | 258592.0 | 0.1 | 0.3 | 0.1 | 65.2 | 523.5 |
| gs250a-2 | c5 | 257920.0 | 0.3 | 258145.0 | 0.1 | 0.3 | 0.1 | 63.3 | 458.3 |
| gs250a-1 | c10 | 258062.0 | 0.3 | 258555.0 | 0.2 | 0.4 | 0.1 | 74.9 | 668.5 |
| gs250a-2 | c10 | 257685.0 | 0.3 | 258223.0 | 0.2 | 0.5 | 0.1 | 60.6 | 341.7 |
| gs250a-1 | c15 | 257843.0 | 0.2 | 258268.0 | 0.2 | 0.3 | 0.1 | 58.9 | 548.6 |
| gs250a-2 | c15 | 257524.0 | 0.2 | 257819.0 | 0.1 | 0.3 | 0.1 | 88.0 | 598.3 |
| gs250a-1 | c20 | 257815.0 | 0.3 | 258239.0 | 0.2 | 0.2 | 0.0 | 87.2 | 598.0 |
| gs250a-2 | c20 | 257452.0 | 0.3 | 257776.0 | 0.1 | 0.3 | 0.1 | 106.0 | 697.4 |
| gs500a-1 | c5 | 511499.0 | 47.9 | 513871.0 | 0.5 | 0.6 | 0.1 | 44.8 | 838.1 |
| gs500a-2 | c5 | 511485.0 | 47.8 | 514124.0 | 0.5 | 0.6 | 0.0 | 55.2 | 845.9 |
| gs500a-1 | c10 | 511056.0 | 30.5 | 513429.0 | 0.5 | 0.5 | 0.0 | 50.4 | 881.1 |
| gs500a-2 | c10 | 511018.0 | 47.9 | 513543.0 | 0.5 | 0.5 | 0.0 | 58.5 | 939.5 |
| gs500a-1 | c15 | 510733.0 | 30.2 | 513165.0 | 0.5 | 0.6 | 0.0 | 48.9 | 928.0 |
| gs500a-2 | c15 | 510718.0 | 31.4 | 513108.0 | 0.5 | 0.6 | 0.1 | 41.7 | 871.2 |
| gs500a-1 | c20 | 510584.0 | 30.9 | 512764.0 | 0.4 | 0.5 | 0.0 | 48.1 | 943.8 |
| gs500a-2 | c20 | 510559.0 | 17.7 | 512560.0 | 0.4 | 0.5 | 0.1 | 51.4 | 906.0 |

of this combination may bring advantages to both approaches: good starting solutions obtained by rounding fractional solutions for the VNS, on one side, and fast high-quality upper bounds for B&C, on the other side.

# References

1. Battiti, R., Tecchiolli, G.: The reactive tabu search. ORSA Journal on Computing 6(2), 126–140 (1994)
2. Ghosh, D.: Neighbourhood search heuristics for the uncapacitated facility location problem. European Journal of Operations Research 150, 150–162 (2003)
3. Gupta, A., Kumar, A., Roughgarden, T.: Simpler and better approximation algorithms for network design. In: STOC, pp. 365–372. ACM, New York (2003)
4. Harm, G., Hentenryck, P.V.: A multistart variable neighborhood search for uncapacitated facility location. In: Proceedings of MIC 2005: The Sixth Metaheuristics International Conference (2005)
5. Hoefer, M.: Experimental comparison of heuristic and approximation algorithms for uncapacitated facility location. In: Jansen, K., Margraf, M., Mastrolli, M., Rolim, J.D.P. (eds.) WEA 2003. LNCS, vol. 2647, pp. 165–178. Springer, Heidelberg (2003)
6. Karger, D.R., Minkoff, M.: Building Steiner trees with incomplete global knowledge. In: FOCS, pp. 613–623 (2000)
7. Khuller, S., Zhu, A.: The general steiner tree-star problem. Information Processing Letters 84(4), 215–220 (2002)
8. Koch, T., Martin, A.: Solving Steiner tree problems in graphs to optimality. Networks 32, 207–232 (1998)
9. Koerkel, M.: On the exact solution of large-scale simple plant location problems. European Journal of Operations Research 39, 157–173 (1989)
10. Kratica, J., Tošić, D., Filipović, V., Ljubić, I.: Solving the simple plant location problem by genetic algorithms. RAIRO - Operations Research 35(1), 127–142 (2001)
11. Lee, Y., Chiu, Y., Ryan, J.: A branch and cut algorithm for a Steiner tree-star problem. INFORMS Journal on Computing 8(3), 194–201 (1996)
12. Ljubić, I.: Exact and Memetic Algorithms for Two Network Design Problems. PhD thesis, Faculty of Computer Science, Vienna University of Technology (November 2004)
13. Ljubić, I., Weiskircher, R., Pferschy, U., Klau, G., Mutzel, P., Fischetti, M.: An algorithmic framework for the exact solution of the prize-collecting Steiner tree problem. Mathematical Progamming, Series B 105(2-3), 427–449 (2006)
14. Mehlhorn, K.: A faster approximation for the Steiner problem in graphs. Information Processing Letters 27, 125–128 (1988)
15. Michel, L., Hentenryck, P.V.: A simple tabu search for warehouse location. European Journal of Operational Research 157(3), 576–591 (2004)
16. Nuggehalli, P., Srinivasan, V., Chiasserini, C.-F.: Energy-efficient caching strategies in ad hoc wireless networks. In: MobiHoc, pp. 25–34 (2003)
17. Swamy, C., Kumar, A.: Primal-dual algorithms for connected facility location problems. Algorithmica 40, 245–269 (2004)
18. Xu, J., Chiu, S.Y., Glover, F.: Using tabu search to solve the Steiner tree-star problem in telecommunications network design. Telecommunication Systems 6(1), 117–125 (1996)

# A Memetic Algorithm for the Optimum Communication Spanning Tree Problem

Thomas Fischer and Peter Merz

Distributed Algorithms Group
University of Kaiserslautern, Germany
{fischer,pmerz}@informatik.uni-kl.de

**Abstract.** For the NP-hard *Optimum Communication Spanning Tree* (OCST) problem a cost minimizing spanning tree has to be found, where the cost depends on the communication volume between each pair of nodes routed over the tree. We present a memetic algorithm (MA) for this problem and focus our discussion on the evaluation of recombination operators for the OCST. The proposed algorithm outperforms evolutionary algorithms (EA) for known benchmark instances and outperforms state-of-the-art solvers for non-Euclidean instances.

## 1  Introduction

The *Optimum Communication Spanning Tree* problem (OCST, a. k. a. minimum communication (cost) spanning tree, MCST) [1] is NP-hard (ND7 in [2]). Given a graph $G = (V, E, d, r)$ with a distance function $d : E \to \mathbb{R}^+$ and a requirement function $r : V \times V \to \mathbb{R}^+$, a spanning tree $T \subseteq G$ is wanted minimizing the cost

$$c(T) = \sum_{i,j \in V} r(i,j) \cdot c(p_{i,j}^T) \tag{1}$$

where $c(p_{i,j}^T)$ is the length of path $p_{i,j}^T$ from $i$ to $j$ in $T$. In the common case, both $d$ and $r$ are symmetric and $d$ does not need to satisfy the triangle inequality.

Our approach is the first memetic algorithm applied to the OCST. It finds solutions for benchmark instances magnitudes faster than pure evolutionary algorithms and operates on instances larger than considered by any other approach. In the remainder of this section related work is discussed. In Section 2 we present the structure of our memetic algorithm. In Section 3 our experimental setup is described. Section 4 discusses the results of the experiments and Section 5 summarizes our findings and suggests directions for future work.

### 1.1  Related Work

Hu [1] defined the OCST, but confined the discussion to two special cases: The *optimum requirement spanning tree* ($d \equiv 1$), which can be solved by an $O(n^4)$ construction algorithm using the Ford-Fulkerson labeling procedure, and the *optimum distance spanning tree* ($r \equiv 1$, a. k. a. minimum routing cost

T. Bartz-Beielstein et al. (Eds.): HM 2007, LNCS 4771, pp. 170–184, 2007.

spanning tree, MRCT). For the latter problem additional requirements were determined which ensure that the optimal solution has a star topology. Other special cases [3] of the OCST are the *product-requirement communication spanning tree* (PROCT, $r(i,j) = r(i) \cdot r(j)$) and the *sum-requirement communication spanning tree* (SROCT, $r(i,j) = r(i) + r(j)$). Approximation algorithms are presented for PROCT ($O(n^5)$ time, 1.577-approximation), SROCT ($O(n^3)$ time, 2-approximation), and MRCT ($O(n^3)$, 1.577-approximation).

Ahuja and Murty [4] define both an exact and an heuristic algorithm for the OCST. For the exact branch & bound algorithm, the lower bound is computed in $O(n^4)$ time and the upper bound is the solution found by the heuristic algorithm. The exact algorithm maintains three sets of edges $I$ (edges included in the spanning tree), $E$ (edges excluded from the tree), and $U$ (undecided edges). Initially, $U$ contains all candidate edges and in each branching step, one edge is removed from $U$ and added to either $I$ or $E$. Details on the heuristic tree construction and improvement algorithms are shown in sections 2.3 and 2.4.

In [5], Palmer and Kershenbaum present a heuristic algorithm which exploits the feature that good solutions for the OCST often have a star topology. Solutions are either constructed by building a star (evaluating all $n$ stars costs $O(n^2)$) or combinations of multiple stars, which have $2 \leq k \ll n$ interior tree nodes (evaluating all $\binom{n}{k}$ stars costs $O(n^{k+1})$). If the star approach is not successful, solutions are created by building minimum spanning trees (MST). A centering node is selected in the tree and the edges get directed towards this network center. The improvement algorithm iteratively evaluates 1-exchange steps by detaching a subtree from its parent and reattaching it to a node between the former parent's parent node and the network center. As an alternative to their heuristic approach, Palmer and Kershenbaum discuss a genetic algorithm [5,6] focusing on the representation of the tree in the GA. Their GA's chromosome contains a vector of bias values for both nodes and edges (*link and node biased*, LNB) modifying the edge cost function. The tree represented by the chromosomes is constructed by building an MST using the modified cost function. The GA performs better than the same authors' heuristic from [5].

The most recent EA for the OCST is from Soak [7]. A new encoding is introduced, which uses a sequence of $2(n-1)$ node ids to represent a spanning tree, where two consecutive nodes form an edge that is included in the tree under construction. Once a cycle is introduced, the longest edge is removed from the cycle and the construction of the tree continues. For recombination, starting from a random node those edges occurring in one of the parents are added to the node sequence. The algorithm finds a new best solution for a 35 node instance from Berry and the known optimum for instance a 100 node instance from Raidl, requiring more than 1 CPU hour for the latter results. For comparison with our findings it should be noted that these results were found with a machine not more than two times slower than our experimental environment.

## 2   Memetic Algorithms

*Evolutionary Algorithms* (EA) [8] are nature-inspired algorithms that try to improve an initial population of solutions by iteratively (in terms of generations) applying a set of operators to the population. Those operators are *recombination*, *mutation* and *selection*. *Local Search* (LS) [9] is a neighborhood search based algorithm. Two solutions are neighbors if one solution can be transformed to the other solution by an elementary modification operation. The effectiveness of an LS depends on the definition of the neighborhood and how the walk through the solution space is performed. *Memetic Algorithms* (MA, Fig. 1) [10] combine both evolutionary algorithms and local search. Using local search inside evolutionary algorithms allows to direct the search process towards better solutions.

```
1  procedure MEMETICALGORITHM
2      P ← INITIALPOPULATION                    ▷ Use some construction heuristic
3      while ¬TERMINATE do          ▷ Termination based on time, convergence, ...
4          P' ← RECOMBINATION(P)       ▷ Create offspring from two or more parents
5          P'' ← MUTATION(P')                    ▷ Perturb offspring solutions
6          P''' ← LOCALSEARCH(P'')             ▷ Optimize offspring solutions
7          P ← SELECTION(P''', P)          ▷ Select individuals for next generation
8      end while
9      return P
10 end procedure
```

**Fig. 1.** Structure of a Memetic Algorithm

Our implementation does not employ a mutation algorithm, as both the local search and the recombination operators may increase diversity. The other components of the memetic algorithm are explained in the remainder of this section and in the presentation of the experimental setup in Sec. 3.

### 2.1   Definitions

A path $p_{a,b}^T$ from $a$ to $b$ in $T$ is a list of nodes $(v_1, \ldots, v_k)$ with $v_1 = a$, $v_k = b$, and $(v_i, v_{i+1}) \in E_T$ for $i = \{1, \ldots, k-1\}$. The length of a path is defined as $c(p_{a,b}^T) = \sum_{i=1}^{k-1} d(v_i, v_{i+1})$. A partial solution $T' \subset T$ is a non-connected graph, the cost $c(T')$ is defined as follows:

$$c(T') = \sum_{i,j \in V} r(i,j) \cdot c'(p_{i,j}^{T'}) \qquad c'(p_{i,j}^{T'}) = \begin{cases} c(p_{i,j}^{T'}) & \text{if } p_{i,j}^{T'} \text{ exists} \\ 0 & \text{otherwise} \end{cases} \qquad (2)$$

In the remainder of this paper the following holds: $n = |V|$ and $m = |E|$.

### 2.2   Common Subroutine

Several algorithms in this paper use a common subroutine ALPHA to estimate the change in cost when performing an 1-exchange move on a tree (edge $e \in E_T$ is removed from $T$ cutting the tree into two components $S$ and $\overline{S}$) or inserting a new edge in the tree during a construction step ($S$ holds the partial solution's nodes, $\overline{S}$ represents all nodes not yet connected to the partial solution). The paths between any pair of nodes within a component have to be passed to ALPHA as parameters $p^S$ and $p^{\overline{S}}$ along with $S$ and $\overline{S}$.

The sum of demands from a node $i$ to all nodes in the other component is denoted by $w_i$. The sum of demands between both components is $\sum_{i \in S} w_i$. A component's external traffic's cost corresponds to $h_i$, if all traffic from $i$'s component to the other component must be routed via $i$.

$$w_i = \begin{cases} \sum_{j \in \overline{S}} r_{i,j} & \text{if } i \in S \\ \sum_{j \in S} r_{i,j} & \text{if } i \in \overline{S} \end{cases} \qquad h_i = \begin{cases} \sum_{j \in S} w_j \cdot c(p^S_{i,j}) & \text{if } i \in S \\ \sum_{j \in \overline{S}} w_j \cdot c(p^{\overline{S}}_{i,j}) & \text{if } i \in \overline{S} \end{cases} \tag{3}$$

Finally, for each edge $(i,j) \in (S \times \overline{S})$, $\alpha_{i,j} = h_i + h_j + d(i,j) \cdot \sum_{i \in S} w_i$ is computed and returned as the subroutine's result. If $S \cup \overline{S} = V$ holds, a tree $T^{(i,j)}$ (both components connected by edge $(i,j)$) has the cost of $c(T^{(i,j)}) = c(T) - \alpha_e + \alpha_{i,j}$. If $S \cup \overline{S} \subsetneq V$, $\alpha_{i,j}$ does not reflect the exact costs and can only be used as a heuristic criterion to select an edge. Using this approach, the cost of evaluating $O(n^2)$ many candidate edges is reduced to $O(n^2)$ time compared to $O(n^3)$ for a naive approach.

### 2.3   Construction of Initial Solutions

**Minimum Spanning Trees.** As argued by Rothlauf et al. in [11], good solutions for the OCST are biased towards minimum spanning trees (MST). The complexity for this algorithm is $O(m + n \log n)$ using Prim's algorithm.

**Star Trees.** Palmer and Kershenbaum suggest in [5] to construct star-shaped trees as initial solutions, motivated by the fact that a star tree is the optimal solution for special cases [1]. The evaluation of all possible trees to find the best star can be done time $O(n^2)$, where the cost of a tree $T_i$ with node $i$ as root is

$$c(T_i) = \sum_{j \in V_T} d(i,j) \cdot w_j \qquad w_j = \sum_{k \in V_T} r(j,k) \tag{4}$$

**Ahuja-Murty Construction.** The tree construction heuristic from [4] closely resembles the local search algorithm from the same paper by starting from a random node and iteratively appending edges to the partial solution. Initially, all communication is routed over the shortest paths in $G$. When adding edges to the partial solution, traffic between nodes in the partial solution has to be routed over the tree increasing the communication cost, therefore the edges to be added are selected to be communication cost optimal. This construction heuristic requires building all-pairs shortest paths $p^G$ in advance, which dominates the total time complexity ($O(n^3)$). The algorithm itself is shown in Fig. 2. The edge evaluation is performed using the subroutine as specified in Sec. 2.2.

### 2.4   Local Search

The local search (Fig. 3) in our algorithm performs a sequence of 1-exchange steps and was originally presented in [12]. Within the local search, all edges

```
1   procedure TREEBUILDING(G(V, E))
2       p^G ← ALLPAIRSSHORTESTPATH(G)              ▷ Building p^G costs O(n^3) time
3       T ← ∅                                      ▷ Empty tree
4       c(p^T) ← 0                                  ▷ No paths in empty tree
5       s ← SEED(G)                                 ▷ Select seeding node
6       S ← {s}
7       S̄ ← V \ S

8       while |S| < |V| do
9           α ← ALPHA(S, S̄, p^G_{i,j}, p^T_{i,j})   ▷ Determine α as described in Sec. 2.2
10          (p, q) ← arg min_{(p,q)∈(S×S̄)}{α_{p,q}} ▷ Select edge with minimal α
11          for each i ∈ S do
12              c(p^T_{i,q}) ← c(p^T_{i,p}) + d(p, q)  ▷ Update path costs for p^T
13          end for
14          S ← S ∪ {q}                             ▷ Update components
15          S̄ ← S̄ \ {q}
16      end while
17      return T
18  end procedure
```

**Fig. 2.** Tree construction heuristic according to Ahuja and Murty [4]

$E_{T^*}$ from the current solution $T^*$ are evaluated in an 1-exchange operation, where the order in which the edges are visited influences the final solution. Our implementation follows the original algorithm's approach to use a queue, which is filled with all tree edges at the beginning (line 3). Edges to be removed are pulled from the queue's head (line 5) and inserted edges are appended to the queue (line 11). Once an edge $e$ has been selected, the tree is cut into two components $S$ and $\overline{S}$ by removing $e$. Whereas the original algorithm evaluates any possible candidate edge from $(S \times \overline{S})$, we select a subset of nodes from both components limiting the set of candidate edges to $(S' \times \overline{S'})$. Various selection strategies have been discussed in [12], however, it has been shown that randomly sampling a small subset of nodes is the best strategy, as any possible candidate edge may be included in $(S' \times \overline{S'})$. In subroutine ALPHA, determining $h$ is the most time consuming part of this algorithm which initially led to the idea of operating on subsets $S' \subseteq S$ and $\overline{S'} \subseteq \overline{S}$.

## 2.5  Recombination

**Exhaustive Recombination.** The *exhaustive recombination algorithm* (Fig. 4) is inspired by the exact algorithm in [4]. The major difference is that only edges from parent trees $T_a$ and $T_b$ are considered in the search. The algorithm begins with two initialization steps. In the first initialization step (lines 2–7) several variables are initialized, such as the initial partial solution $T'$, containing only edges common to both parent trees, and edge sets $E^*$ and $E'$ containing edges occurring only in the better or worse parent, respectively. The second initialization step (lines 8–13) builds a stack of partial solutions and candidate edge set starting from the partial solution $T'$ containing only common edges. Iteratively,

```
 1  procedure LocalSearch(T)
 2      T* ← T                                      ▷ T* is current favorite solution
 3      Q ← E_T                                      ▷ Initialize queue
 4      while not every edge in Q visited do
 5          e = Pull(Q)                              ▷ Fetch edge from queue
 6          (S, S̄) ← CutTree(T*, e)        ▷ Cut T* into two components S and S̄
 7          (S', S̄') ← SelectCities(S, S̄)              ▷ S' ⊆ S, S̄' ⊆ S̄
 8          α ← Alpha(S', S̄', p_{i,j}^{T*}, p_{i,j}^{T*})    ▷ Determine α as described in Sec. 2.2
 9          (p, q) ← arg min_{(p,q)∈(S'×S̄')}{α_{p,q}}       ▷ Select edge with minimal α
10          T* ← T* \ {e} ∪ {(p, q)}                      ▷ Update tree T*
11          Push(Q, (p, q))        ▷ Add edge to queue       ▷ Add new edge to queue
12      end while
13      return T*                                      ▷ Return best solution
14  end procedure
```

**Fig. 3.** Iterated Local Search (1-exchange neighborhood) with Random Sampling

edges from $E^*$ are added to the tree which is pushed on the stack together with the edge set $E'$. This strategy fills the branch & bound-like algorithm's stack as if the algorithm had always branched by selecting edges from the better parent solution. Thus the main loop (lines 14–32) starts with partial solutions close to the better solution parent solution $T_a$, but departs from it when exploring the search space further. At the beginning of each loop iteration, a position in the search space consisting of a partial solution $T'$ and a set of candidate edges $E'$ is fetched from the stack. If the partial solution $T'$ is actually a valid solution, it may be stored as a candidate final solution and the upper bound is lowered. Otherwise, an edge $e$ is selected. The algorithm branches in two possible search space regions containing trees with or without $e$, respectively. To increase the efficiency of the search, the partial solution and the candidate edge set are checked if they combined hold enough edges for future valid solutions (line 24) and if the partial solution including $e$ does not contain a cycle and is below the upper bound (line 28). For each passed check, the modified partial solution and candidate edge sets are pushed on the stack. The main loop terminates when the stack is empty, but may terminate earlier once a given time bound or iteration count is reached.

**Tree Building Recombination.** The *tree building recombination algorithm* (Fig. 5) reuses concepts from the tree building algorithm from [4]. This recombination operator determines the set of common fragments $\mathcal{T}$ in both parents $T_a$ and $T_b$ first. A fragment is defined as a connected set of edges that are contained in both parents. A path length matrix $c(p_{i,j}^{\mathcal{T}})$ is initialized with the all-pairs shortest paths' length in $G$ and updated with the unique paths' lengths from the fragments. The offspring solution is initialized by the fragment that contains a randomly selected node $s$ (line 5). As long as not all fragments are connected in the offspring, edge $(p, q) \in S \times \overline{S}$ which increases the offspring cost the least is selected for insertion evaluating the set of fragments in Alpha (see Sec. 2.2). Once $q$ has been determined, its fragment is inserted into the partial tree and the path length matrix is updated with the new edges in the partial solution.

```
 1  procedure EXHAUSTIVERECOMBINATION(T_a, T_b)
 2      S ← ∅                                          ▷ Start with empty stack
 3      T* ← T_a                                        ▷ Initial best solution
 4      z* ← c(T_a)                                     ▷ Set upper bound
 5      T' ← E_{T_a} ∩ E_{T_b}                          ▷ Initial partial solution with common edges
 6      E* ← E_{T_a} \ E_{T_b}                          ▷ Edges only in better parent
 7      E' ← E_{T_b} \ E_{T_a}                          ▷ Edges only in worse parent

 8      while |E*| > 0 do                               ▷ Prepare stack
 9          PUSH(S, (T', E'))                           ▷ Put current state on stack
10          e ← arg min_{e∈E*} d(e)                     ▷ Fetch shortest edge from E*
11          T' ← T' ∪ {e}                               ▷ Add e to growing partial solution
12          E* ← E* \ {e}
13      end while

14      while |S| > 0 do
15          (T', E') ← POP(S)                           ▷ Fetch current state from stack
16          if ISTREE(T') then                          ▷ Valid solution found
17              if c(T') < c(T*) then                   ▷ New best tree found
18                  T* ← T'                             ▷ Store new best tree
19                  z* ← c(T*)                          ▷ Update upper bound
20              end if
21          else if |E'| > 0 then                       ▷ Branch search
22              e ← arg min_{e∈E'} d(e)                 ▷ Fetch shortest edge from E'
23              E' ← E' \ {e}
24              if |E'| + |E_{T'}| ≥ |V| − 1 then       ▷ Enough edges left for a tree?
25                  PUSH(S, (T', E'))                   ▷ Branch that does not contain edge e
26              end if
27              T' ← T' ∪ {e}
28              if ISVALID(T') ∧ c'(T') < z* then       ▷ T' still valid and below LB?
29                  PUSH(S, (T', E'))                   ▷ Branch that must contain edge e
30              end if
31          end if
32      end while
33      return T*                                       ▷ Return best solution
34  end procedure
```

**Fig. 4.** Recombination of two parent trees $T_a$ and $T_b$ to an offspring tree $T^*$ using the Exhaustive recombination algorithm. W. l. o g. $c(T_a) \leq c(T_b)$

**Path Merging Recombination.** The *path merging recombination algorithm* (Fig. 6) is the only operator that does not use the OCST cost function during recombination. Starting with a random seeding node in $S$ and $\overline{S} = V \setminus S$, iteratively, until all nodes are connected by the partial solution, a node $s \in \overline{S}$ and one parent $T \in \{T_a, T_b\}$ are selected (lines 7 and 8). Using Dijkstra's shortest path algorithm, the shortest path in $T$ from $s$ to the closest node in $S$ is determined. All edges from this path are added to the partial tree (line 10), nodes from the path are added to $S$. No cycles can occur.

```
1  procedure TREEBUILDINGRECOMBINATION(T_a, T_b)
2      𝒯 ← {T_1, ..., T_k}                          ▷ Determine fragments of common edges
3      p^𝒯 ← ALLPAIRSSHORTESTPATH(G)                ▷ Building p^G costs O(n^3) time
4      s ← SEED(G)                                   ▷ Select seeding node
5      T* ← arg_{s∈T} T ∈ 𝒯                         ▷ Initialize partial solution with s's fragment
6      𝒯 ← 𝒯 \ {T*}                                 ▷ Update fragment set
7      S ← V_{T*}                                    ▷ Update components
8      S̄ ← V \ S
9      while |S| < |V| do
10         α ← ALPHA(S, S̄, p^𝒯_{i,j}, p^𝒯_{i,j})     ▷ Determine α as described in Sec. 2.2
11         (p, q) ← arg min_{(p,q)∈(S×S̄)}{α_{p,q}}   ▷ Select edge with minimal α
12         T' ← arg_{q∈T} T ∈ 𝒯                      ▷ Determine q's fragment
13         𝒯 ← 𝒯 \ {T'}                              ▷ Update fragment set
14         T* ← T* ∪ T'                              ▷ Add q's fragment to partial solution
15         for each (i, j) ∈ (S × V_{T'}) do
16             c(p^𝒯_{i,j}) ← c(p^𝒯_{i,p}) + d(p,q) + c(p^𝒯_{q,j})    ▷ Update path costs for p^𝒯_{i,j}
17         end for
18         S ← S ∪ V_{T'}                            ▷ Update components
19         S̄ ← V \ S
20     end while
21     return T*                                     ▷ Return solution
22 end procedure
```

**Fig. 5.** Recombination of two parent trees $T_a$ and $T_b$ to an offspring tree $T^*$ using the Tree Building recombination algorithm

```
1  procedure PATHRECOMBINATION(T_a, T_b)
2      T* ← ∅
3      s ← SEED(G)                                   ▷ Select seeding node
4      S ← {s}                                       ▷ Initialize components
5      S̄ ← V \ S
6      while |S| < |V| do
7          s ← SEED(S̄)                              ▷ Select node not in S
8          T ← SELECT(T_a, T_b)                      ▷ Select tree, alternating between T_a and T_b
9          p ← arg min_{p^T_{s,s'} with s'∈S} |p^T_{s,s'}|    ▷ Find shortest path in G to a node in S
10         T* ← T* ∪ p                               ▷ Add path to partial solution
11         S ← S ∪ V_p                               ▷ Update components
12         S̄ ← S̄ \ V_p
13     end while
14     return T*                                     ▷ Return solution
15 end procedure
```

**Fig. 6.** Recombination of two parent trees $T_a$ and $T_b$ to an offspring tree $T^*$ using the Path recombination algorithm

## 3   Experimental Setup

We used both standard benchmark instances and random problem instances constructed similar to the specification in [13]. From the set of benchmark instances, we selected two instances from the Raidl series (50 and 100 nodes). The randomly constructed problem instances use a two-dimensional grid of size $1000 \times 1000$. Demands between nodes are randomly selected from the interval $[0, 100]$. For instances where non-Euclidean distances were used, the distances were taken from the interval $[0, 100]$. For Euclidean instances, node coordinates are randomly selected and the distance from node $i$ to $j$ is defined as $d(i, j) = \lfloor \sqrt{(x_i - x_j)^2 + (y_i - y_j)^2} + 0.5 \rfloor$. Coordinates, distances, and demands are restricted to natural numbers. Here, four random instances were created with sizes 300 and 500 using both Euclidean and non-Euclidean distances. The best known solutions are 806 864 (`Raidl.50`), 2 561 543 (`Raidl.100`), 15 330 782 (`Rand.300`), 1 612 822 306 (`Rand.300.E`), 37 744 037 (`Rand.500`), and 4 481 969 584 (`Rand.500.E`).

Initial trees were constructed using either the construction heuristic from Ahuja and Murty (AM-C), a minimum spanning tree (MST), a random tree (RAND) or using the best star tree (STAR). These construction heuristic were randomly seeded, if possible. The population size of the MA was set to either 2, 4, or 8. Recombination operators from Sec. 2.5 were used: Tree-building (TB), Path Merging (PM), and Exhaustive recombination (EXH). For comparison, a Replacement recombination (RPL) was also applied using the better of two parents as the new offspring. Each individual of the current population is guaranteed to be part of at least one recombination. Offsprings from the recombination were subject to a local search improvement as described in Sec. 2.4. No explicit mutation was performed. Iteratively, until the next population was complete, two individuals were randomly selected from the combined set of parent and offspring individuals and the better of both was inserted into the next generation.

For comparison, each instance and each initial solution was also solved by iterated local search algorithms (ILS) using our implementation of the Ahuja and Murty tree improvement heuristic (AM-H) and the Random Sampling AM algorithm (RSAM) as described in [12]. For an instance with $n$ nodes the termination criterion was the time limit of $\lfloor n/50 \rfloor^2$ CPU seconds. Alternatively, a convergence detection heuristic terminated an algorithm once it did not find any improvement within the last half of its elapsed running time. Experiments were conducted on a 3.0 GHz Pentium 4 CPU. Every setup was repeated 15 times, average values were used for discussion.

## 4   Results

Results as discussed in this section are shown in Tables 1, 2, and 3. The first three columns of each table describe the setup, the other columns are ordered into four groups of two columns each. Each column group shows the average cost and excess of setups when starting with one of the four construction heuristic.

For each instance, the first four rows summarize the initial solution cost, the best known solution, and the results for both iterated local search algorithms AM-H and RSAM. The remaining rows for each instance are ordered into four groups of rows summarizing the results from each of the four recombination operators, each having three rows for setups with population size 2, 4, or 8.

**Table 1.** Results for the instances from the Raidl series (50 or 100 nodes, respectively)

| Instance | Group | Pop | Constr. Heur. → RAND | | MST | | STAR | | AM-C | |
|---|---|---|---|---|---|---|---|---|---|---|
| | | ↓ Solver | Cost | Excess | Cost | Excess | Cost | Excess | Cost | Excess |
| Raidl.50 | | Init | $24.0 \cdot 10^6$ | 2879 % | 979341 | 21.38 % | 4372492 | 442 % | 813948 | 0.88 % |
| | | Best | 806864 | 0.00 % | 806864 | 0.00 % | 806864 | 0.00 % | 806864 | 0.00 % |
| | | AM-H | 807244 | 0.05 % | 807516 | 0.08 % | 806972 | 0.01 % | 808061 | 0.15 % |
| | | RSAM | 816103 | 1.14 % | 814725 | 0.97 % | 815607 | 1.08 % | 811048 | 0.52 % |
| | TB | 2 | 807131 | 0.03 % | 807029 | 0.02 % | 806864 | 0.00 % | 807027 | 0.02 % |
| | | 4 | 806864 | 0.00 % | 807079 | 0.03 % | 806891 | 0.00 % | 806864 | 0.00 % |
| | | 8 | 806864 | 0.00 % | 807289 | 0.05 % | 806864 | 0.00 % | 806864 | 0.00 % |
| | EXH | 2 | $21.5 \cdot 10^6$ | 2573 % | 807217 | 0.04 % | 876360 | 8.61 % | 806864 | 0.00 % |
| | | 4 | $19.5 \cdot 10^6$ | 2320 % | 807638 | 0.10 % | 1113753 | 38.03 % | 806864 | 0.00 % |
| | | 8 | $18.7 \cdot 10^6$ | 2228 % | 807599 | 0.09 % | 1377920 | 70.77 % | 806864 | 0.00 % |
| | PM | 2 | 807615 | 0.09 % | 806864 | 0.00 % | 807083 | 0.03 % | 806864 | 0.00 % |
| | | 4 | 807371 | 0.06 % | 807398 | 0.07 % | 807500 | 0.08 % | 806864 | 0.00 % |
| | | 8 | 814368 | 0.93 % | 810679 | 0.47 % | 810520 | 0.45 % | 806864 | 0.00 % |
| | RPL | 2 | 806889 | 0.00 % | 806889 | 0.00 % | 806864 | 0.00 % | 806864 | 0.00 % |
| | | 4 | 806889 | 0.00 % | 807487 | 0.08 % | 806864 | 0.00 % | 806864 | 0.00 % |
| | | 8 | 807425 | 0.07 % | 807221 | 0.04 % | 806864 | 0.00 % | 806864 | 0.00 % |
| Raidl.100 | | Init | $143.6 \cdot 10^6$ | 5507 % | 3486328 | 36.10 % | $20.6 \cdot 10^6$ | 708 % | 2644050 | 3.22 % |
| | | Best | 2561543 | 0.00 % | 2561543 | 0.00 % | 2561543 | 0.00 % | 2561543 | 0.00 % |
| | | AM-H | 2661344 | 3.90 % | 2564279 | 0.11 % | 2600212 | 1.51 % | 2639093 | 3.03 % |
| | | RSAM | 2573098 | 0.45 % | 2574815 | 0.52 % | 2574302 | 0.50 % | 2574172 | 0.49 % |
| | TB | 2 | 2564878 | 0.13 % | 2562310 | 0.03 % | 2601058 | 1.54 % | 2589910 | 1.11 % |
| | | 4 | 2564511 | 0.12 % | 2561909 | 0.01 % | 2575482 | 0.54 % | 2566027 | 0.17 % |
| | | 8 | 2562643 | 0.04 % | 2561771 | 0.01 % | 2574373 | 0.50 % | 2564389 | 0.11 % |
| | EXH | 2 | $132.7 \cdot 10^6$ | 5082 % | 2562276 | 0.03 % | 9376789 | 266 % | 2595384 | 1.32 % |
| | | 4 | $121.0 \cdot 10^6$ | 4627 % | 2610051 | 1.89 % | 9558582 | 273 % | 2569575 | 0.31 % |
| | | 8 | $110.2 \cdot 10^6$ | 4205 % | 2614602 | 2.07 % | 9525772 | 272 % | 2568730 | 0.28 % |
| | PM | 2 | 2649386 | 3.43 % | 2575410 | 0.54 % | 2609416 | 1.87 % | 2589611 | 1.10 % |
| | | 4 | 2644876 | 3.25 % | 2563409 | 0.07 % | 2667741 | 4.15 % | 2564360 | 0.11 % |
| | | 8 | 2707643 | 5.70 % | 2585404 | 0.93 % | 2702280 | 5.49 % | 2565865 | 0.17 % |
| | RPL | 2 | 2660527 | 3.86 % | 2574575 | 0.51 % | 2639802 | 3.06 % | 2586628 | 0.98 % |
| | | 4 | 2635336 | 2.88 % | 2562625 | 0.04 % | 2650752 | 3.48 % | 2562574 | 0.04 % |
| | | 8 | 2683333 | 4.75 % | 2561543 | 0.00 % | 2650826 | 3.49 % | 2561543 | 0.00 % |

**Tree Construction.** Regarding the quality of the initial solutions, the construction heuristic AM-C finds the best trees for the Raidl series instances. For `Raidl.50`, it finds the optimal tree in about half of all cases resulting in an average excess of 0.88 %. The second best construction heuristic is MST, finding trees with an excess of 21.4 % (`Raidl.50`) to 36.1 % (`Raidl.100`). Star trees are surprisingly poor solutions resulting in an excess of 442 % (`Raidl.50`) to 708 % (`Raidl.100`). For random instances, construction heuristic show different performance depending on whether the instance uses Euclidean or random distances. For Euclidean trees, star trees represent the best initial solution (4.91 % excess for `Rand.500.E`), whereas for random distances, AM-C performs best (3.62 % excess for `Rand.500`). For both instance types, MST is the second best choice

**Table 2.** Results for random instances with 300 nodes, either without or with Euclidean distances (upper and lower table half, respectively).

| Instance | Constr. Heur. → ↓ Solver | | RAND Cost | RAND Excess | MST Cost | MST Excess | STAR Cost | STAR Excess | AM-C Cost | AM-C Excess |
|---|---|---|---|---|---|---|---|---|---|---|
| Rand.300 | Init | | $2.3 \cdot 10^9$ | 15202 % | $25.0 \cdot 10^6$ | 63.18 % | $202.6 \cdot 10^6$ | 1222 % | $15.7 \cdot 10^6$ | 3.05 % |
| Rand.300 | Best | | $15.3 \cdot 10^6$ | 0.00 % | $15.3 \cdot 10^6$ | 0.00 % | $15.3 \cdot 10^6$ | 0.00 % | $15.3 \cdot 10^6$ | 0.00 % |
| Rand.300 | AM-H | | $15.9 \cdot 10^6$ | 3.88 % | $15.8 \cdot 10^6$ | 3.48 % | $15.7 \cdot 10^6$ | 2.46 % | $15.5 \cdot 10^6$ | 1.67 % |
| Rand.300 | RSAM | | $15.5 \cdot 10^6$ | 1.63 % | $15.7 \cdot 10^6$ | 2.48 % | $15.7 \cdot 10^6$ | 2.65 % | $15.6 \cdot 10^6$ | 1.79 % |
| Rand.300 | TB | 2 | $15.4 \cdot 10^6$ | 0.77 % | $15.6 \cdot 10^6$ | 1.81 % | $15.7 \cdot 10^6$ | 2.49 % | $15.4 \cdot 10^6$ | 0.89 % |
| Rand.300 | TB | 4 | $15.3 \cdot 10^6$ | 0.44 % | $15.3 \cdot 10^6$ | 0.33 % | $15.7 \cdot 10^6$ | 2.44 % | $15.3 \cdot 10^6$ | 0.42 % |
| Rand.300 | TB | 8 | $15.3 \cdot 10^6$ | 0.06 % | $15.3 \cdot 10^6$ | 0.13 % | $15.7 \cdot 10^6$ | 3.05 % | $15.3 \cdot 10^6$ | 0.12 % |
| Rand.300 | EXH | 2 | $2.1 \cdot 10^9$ | 13724 % | $23.8 \cdot 10^6$ | 55.66 % | $53.1 \cdot 10^6$ | 247 % | $15.5 \cdot 10^6$ | 1.17 % |
| Rand.300 | EXH | 4 | $2.0 \cdot 10^9$ | 13198 % | $22.9 \cdot 10^6$ | 50.00 % | $51.3 \cdot 10^6$ | 235 % | $15.4 \cdot 10^6$ | 0.52 % |
| Rand.300 | EXH | 8 | $1.9 \cdot 10^9$ | 12352 % | $22.5 \cdot 10^6$ | 46.92 % | $51.0 \cdot 10^6$ | 233 % | $15.3 \cdot 10^6$ | 0.23 % |
| Rand.300 | PM | 2 | $15.5 \cdot 10^6$ | 1.58 % | $15.6 \cdot 10^6$ | 2.25 % | $15.7 \cdot 10^6$ | 2.84 % | $15.5 \cdot 10^6$ | 1.14 % |
| Rand.300 | PM | 4 | $15.9 \cdot 10^6$ | 4.22 % | $15.7 \cdot 10^6$ | 2.81 % | $15.8 \cdot 10^6$ | 3.38 % | $15.4 \cdot 10^6$ | 0.52 % |
| Rand.300 | PM | 8 | $16.5 \cdot 10^6$ | 7.68 % | $16.1 \cdot 10^6$ | 5.27 % | $16.4 \cdot 10^6$ | 7.51 % | $15.3 \cdot 10^6$ | 0.23 % |
| Rand.300 | RPL | 2 | $15.7 \cdot 10^6$ | 2.77 % | $16.0 \cdot 10^6$ | 4.65 % | $15.6 \cdot 10^6$ | 2.10 % | $15.4 \cdot 10^6$ | 0.79 % |
| Rand.300 | RPL | 4 | $15.8 \cdot 10^6$ | 3.25 % | $15.7 \cdot 10^6$ | 2.47 % | $15.7 \cdot 10^6$ | 2.44 % | $15.3 \cdot 10^6$ | 0.37 % |
| Rand.300 | RPL | 8 | $15.8 \cdot 10^6$ | 3.26 % | $15.6 \cdot 10^6$ | 2.19 % | $15.8 \cdot 10^6$ | 3.21 % | $15.3 \cdot 10^6$ | 0.07 % |
| Rand.300.E | Init | | $24.2 \cdot 10^9$ | 1406 % | $2.7 \cdot 10^9$ | 68.09 % | $1.6 \cdot 10^9$ | 5.19 % | $5.4 \cdot 10^9$ | 239 % |
| Rand.300.E | Best | | $1.6 \cdot 10^9$ | 0.00 % | $1.6 \cdot 10^9$ | 0.00 % | $1.6 \cdot 10^9$ | 0.00 % | $1.6 \cdot 10^9$ | 0.00 % |
| Rand.300.E | AM-H | | $1.6 \cdot 10^9$ | 1.71 % | $1.6 \cdot 10^9$ | 1.18 % | $1.6 \cdot 10^9$ | 0.47 % | $1.6 \cdot 10^9$ | 0.71 % |
| Rand.300.E | RSAM | | $1.6 \cdot 10^9$ | 0.72 % | $1.6 \cdot 10^9$ | 0.76 % | $1.6 \cdot 10^9$ | 0.65 % | $1.6 \cdot 10^9$ | 0.74 % |
| Rand.300.E | TB | 2 | $1.6 \cdot 10^9$ | 1.90 % | $1.6 \cdot 10^9$ | 1.73 % | $1.6 \cdot 10^9$ | 1.62 % | $1.6 \cdot 10^9$ | 1.88 % |
| Rand.300.E | TB | 4 | $1.6 \cdot 10^9$ | 2.89 % | $1.7 \cdot 10^9$ | 6.70 % | $1.6 \cdot 10^9$ | 2.52 % | $1.7 \cdot 10^9$ | 5.61 % |
| Rand.300.E | TB | 8 | $1.6 \cdot 10^9$ | 4.74 % | $1.6 \cdot 10^9$ | 4.20 % | $1.6 \cdot 10^9$ | 3.13 % | $1.6 \cdot 10^9$ | 4.24 % |
| Rand.300.E | EXH | 2 | $22.0 \cdot 10^9$ | 1269 % | $1.8 \cdot 10^9$ | 17.25 % | $1.6 \cdot 10^9$ | 5.06 % | $5.2 \cdot 10^9$ | 228 % |
| Rand.300.E | EXH | 4 | $20.7 \cdot 10^9$ | 1187 % | $1.8 \cdot 10^9$ | 13.19 % | $1.6 \cdot 10^9$ | 4.98 % | $5.2 \cdot 10^9$ | 228 % |
| Rand.300.E | EXH | 8 | $20.0 \cdot 10^9$ | 1140 % | $1.8 \cdot 10^9$ | 12.41 % | $1.6 \cdot 10^9$ | 5.09 % | $5.2 \cdot 10^9$ | 223 % |
| Rand.300.E | PM | 2 | $1.6 \cdot 10^9$ | 0.87 % | $1.6 \cdot 10^9$ | 0.78 % | $1.6 \cdot 10^9$ | 0.68 % | $1.6 \cdot 10^9$ | 0.77 % |
| Rand.300.E | PM | 4 | $1.6 \cdot 10^9$ | 1.10 % | $1.6 \cdot 10^9$ | 1.18 % | $1.6 \cdot 10^9$ | 1.02 % | $1.6 \cdot 10^9$ | 1.09 % |
| Rand.300.E | PM | 8 | $1.6 \cdot 10^9$ | 1.59 % | $1.6 \cdot 10^9$ | 1.58 % | $1.6 \cdot 10^9$ | 1.32 % | $1.6 \cdot 10^9$ | 1.85 % |
| Rand.300.E | RPL | 2 | $1.6 \cdot 10^9$ | 1.04 % | $1.6 \cdot 10^9$ | 0.97 % | $1.6 \cdot 10^9$ | 0.45 % | $1.6 \cdot 10^9$ | 0.79 % |
| Rand.300.E | RPL | 4 | $1.6 \cdot 10^9$ | 0.95 % | $1.6 \cdot 10^9$ | 0.93 % | $1.6 \cdot 10^9$ | 0.56 % | $1.6 \cdot 10^9$ | 0.74 % |
| Rand.300.E | RPL | 8 | $1.6 \cdot 10^9$ | 0.86 % | $1.6 \cdot 10^9$ | 1.07 % | $1.6 \cdot 10^9$ | 0.74 % | $1.6 \cdot 10^9$ | 0.81 % |

with an excess of 68.9 % and 102 %, respectively. Finally, random trees are the worst choice in any case having an excess of >1000 %.

**Local Search.** For small instances, the AM-H performs better compared to RSAM. E. g. AM-H finds solutions with excess 0.05 % for `Raidl.50` starting from random trees, whereas RSAM's average excess is 1.14 %. For larger instances, RSAM usually performs better, as for the same setup with `Raidl.100` the RSAM's average excess is 0.45 % compared to 3.90 % for AM-H. For non-Euclidean random instances, AM-H performs best when starting from AM-C-based trees. For these instances, the average cost depends more on the initial tree compared to RSAM. E. g. for `Rand.300`, the excess ranges from 1.67 % (AM-C) to 3.88 % (RAND), whereas RSAM's excess ranges from 1.63 % (RAND) to 2.65 % (STAR) and is thus more independent from the initial tree. For Euclidean random instances, best results can be expected starting from star trees

**Table 3.** Results for random instances with 500 nodes, either without or with Euclidean distances (upper and lower table half, respectively).

| Instance | Constr. Heur. → ↓ Solver | | RAND Cost | RAND Excess | MST Cost | MST Excess | STAR Cost | STAR Excess | AM-C Cost | AM-C Excess |
|---|---|---|---|---|---|---|---|---|---|---|
| Rand.500 | Init | | $8.4 \cdot 10^9$ | 22336 % | $63.7 \cdot 10^6$ | 68.88 % | $582.6 \cdot 10^6$ | 1444 % | $39.1 \cdot 10^6$ | 3.62 % |
| | Best | | $37.7 \cdot 10^6$ | 0.00 % | $37.7 \cdot 10^6$ | 0.00 % | $37.7 \cdot 10^6$ | 0.00 % | $37.7 \cdot 10^6$ | 0.00 % |
| | AM-H | | $40.2 \cdot 10^6$ | 6.63 % | $40.0 \cdot 10^6$ | 6.08 % | $41.8 \cdot 10^6$ | 10.95 % | $38.8 \cdot 10^6$ | 2.91 % |
| | RSAM | | $39.2 \cdot 10^6$ | 3.88 % | $38.8 \cdot 10^6$ | 3.03 % | $40.6 \cdot 10^6$ | 7.73 % | $38.8 \cdot 10^6$ | 2.95 % |
| | TB | 2 | $38.4 \cdot 10^6$ | 1.88 % | $39.6 \cdot 10^6$ | 5.09 % | $41.3 \cdot 10^6$ | 9.65 % | $38.3 \cdot 10^6$ | 1.69 % |
| | | 4 | $38.1 \cdot 10^6$ | 1.06 % | $38.5 \cdot 10^6$ | 2.15 % | $40.5 \cdot 10^6$ | 7.33 % | $38.1 \cdot 10^6$ | 1.15 % |
| | | 8 | $37.9 \cdot 10^6$ | 0.52 % | $38.7 \cdot 10^6$ | 2.59 % | $40.5 \cdot 10^6$ | 7.34 % | $37.9 \cdot 10^6$ | 0.61 % |
| | EXH | 2 | $8.0 \cdot 10^9$ | 21342 % | $61.1 \cdot 10^6$ | 61.92 % | $117.3 \cdot 10^6$ | 211 % | $38.4 \cdot 10^6$ | 1.95 % |
| | | 4 | $7.2 \cdot 10^9$ | 19223 % | $58.5 \cdot 10^6$ | 55.25 % | $114.7 \cdot 10^6$ | 204 % | $38.2 \cdot 10^6$ | 1.23 % |
| | | 8 | $6.8 \cdot 10^9$ | 18136 % | $56.5 \cdot 10^6$ | 49.74 % | $115.7 \cdot 10^6$ | 207 % | $38.0 \cdot 10^6$ | 0.71 % |
| | PM | 2 | $39.5 \cdot 10^6$ | 4.91 % | $39.8 \cdot 10^6$ | 5.53 % | $40.7 \cdot 10^6$ | 8.09 % | $38.4 \cdot 10^6$ | 1.95 % |
| | | 4 | $40.0 \cdot 10^6$ | 6.16 % | $40.0 \cdot 10^6$ | 6.12 % | $41.3 \cdot 10^6$ | 9.59 % | $38.2 \cdot 10^6$ | 1.23 % |
| | | 8 | $42.1 \cdot 10^6$ | 11.72 % | $53.2 \cdot 10^6$ | 40.98 % | $43.0 \cdot 10^6$ | 14.13 % | $38.0 \cdot 10^6$ | 0.71 % |
| | RPL | 2 | $40.2 \cdot 10^6$ | 6.63 % | $39.8 \cdot 10^6$ | 5.53 % | $41.0 \cdot 10^6$ | 8.68 % | $38.3 \cdot 10^6$ | 1.62 % |
| | | 4 | $40.1 \cdot 10^6$ | 6.34 % | $39.6 \cdot 10^6$ | 5.03 % | $41.0 \cdot 10^6$ | 8.71 % | $38.1 \cdot 10^6$ | 1.01 % |
| | | 8 | $40.0 \cdot 10^6$ | 6.17 % | $39.5 \cdot 10^6$ | 4.91 % | $40.9 \cdot 10^6$ | 8.59 % | $37.9 \cdot 10^6$ | 0.57 % |
| Rand.500.E | Init | | $86.0 \cdot 10^9$ | 1821 % | $9.0 \cdot 10^9$ | 102 % | $4.7 \cdot 10^9$ | 4.91 % | $16.5 \cdot 10^9$ | 270 % |
| | Best | | $4.4 \cdot 10^9$ | 0.00 % | $4.4 \cdot 10^9$ | 0.00 % | $4.4 \cdot 10^9$ | 0.00 % | $4.4 \cdot 10^9$ | 0.00 % |
| | AM-H | | $4.5 \cdot 10^9$ | 1.31 % | $4.5 \cdot 10^9$ | 1.35 % | $4.5 \cdot 10^9$ | 0.58 % | $4.5 \cdot 10^9$ | 1.02 % |
| | RSAM | | $4.5 \cdot 10^9$ | 0.53 % | $4.5 \cdot 10^9$ | 0.68 % | $4.4 \cdot 10^9$ | 0.32 % | $4.5 \cdot 10^9$ | 0.56 % |
| | TB | 2 | $4.8 \cdot 10^9$ | 7.11 % | $4.6 \cdot 10^9$ | 3.60 % | $4.6 \cdot 10^9$ | 4.39 % | $4.6 \cdot 10^9$ | 4.39 % |
| | | 4 | $4.8 \cdot 10^9$ | 7.58 % | $4.8 \cdot 10^9$ | 7.58 % | $4.6 \cdot 10^9$ | 4.75 % | $4.7 \cdot 10^9$ | 7.00 % |
| | | 8 | $4.7 \cdot 10^9$ | 6.29 % | $4.7 \cdot 10^9$ | 5.61 % | $4.6 \cdot 10^9$ | 4.44 % | $15.4 \cdot 10^9$ | 244 % |
| | EXH | 2 | $80.6 \cdot 10^9$ | 1699 % | $4.9 \cdot 10^9$ | 9.33 % | $4.7 \cdot 10^9$ | 4.87 % | $16.0 \cdot 10^9$ | 258 % |
| | | 4 | $73.3 \cdot 10^9$ | 1536 % | $4.8 \cdot 10^9$ | 7.81 % | $4.7 \cdot 10^9$ | 4.87 % | $15.8 \cdot 10^9$ | 253 % |
| | | 8 | $68.9 \cdot 10^9$ | 1439 % | $4.8 \cdot 10^9$ | 7.44 % | $4.7 \cdot 10^9$ | 4.87 % | $15.5 \cdot 10^9$ | 248 % |
| | PM | 2 | $4.5 \cdot 10^9$ | 0.89 % | $4.5 \cdot 10^9$ | 0.71 % | $4.5 \cdot 10^9$ | 0.59 % | $4.5 \cdot 10^9$ | 0.86 % |
| | | 4 | $4.5 \cdot 10^9$ | 1.04 % | $4.5 \cdot 10^9$ | 0.98 % | $4.5 \cdot 10^9$ | 0.80 % | $4.5 \cdot 10^9$ | 1.17 % |
| | | 8 | $4.5 \cdot 10^9$ | 1.43 % | $4.5 \cdot 10^9$ | 1.44 % | $4.5 \cdot 10^9$ | 1.20 % | $4.5 \cdot 10^9$ | 2.29 % |
| | RPL | 2 | $4.5 \cdot 10^9$ | 0.98 % | $4.5 \cdot 10^9$ | 0.73 % | $4.5 \cdot 10^9$ | 0.42 % | $4.5 \cdot 10^9$ | 0.85 % |
| | | 4 | $4.5 \cdot 10^9$ | 0.95 % | $4.5 \cdot 10^9$ | 0.73 % | $4.5 \cdot 10^9$ | 0.48 % | $4.5 \cdot 10^9$ | 0.90 % |
| | | 8 | $4.5 \cdot 10^9$ | 0.99 % | $4.5 \cdot 10^9$ | 0.81 % | $4.5 \cdot 10^9$ | 0.63 % | $4.5 \cdot 10^9$ | 1.33 % |

when using AM-H. E. g. for `Rand.500.E`, trees with an excess of 0.58 % were found compared to solutions based on initial solutions from AM-C (1.02 %) or MST (1.35 %). Again, the larger the instance, the better RSAM performs compared to AM-H confirming our findings from [12]. E. g. for `Rand.500`, the RSAM approach finds better solutions in three out of four setups.

**Tree Building Recombination Operator.** For Raidl and random non-Euclidean instances, the TB operator is the fastest converging recombinator when starting from random trees. For the other types of initial trees, this recombination operator is performing well, too, but for several setups RPL and PM perform better. Larger populations usually allow this recombinator to find good solutions even faster. E. g. for `Rand.300` starting from an MST, the average excess is 1.81 % for a population of size 2, but it is 0.13 % for populations of size 8. For Euclidean distance based instances, however, the same recombination operator is surpassed by all other setups except EXH. For instances where AM-C finds good initial trees, TB performs very well starting from random solutions, as in this case the two parent trees have few edges in common and thus TB rebuilds the tree using an AM-C-like heuristic resulting in much better offsprings. Drawback of TB is that it becomes too expensive for larger instances due to its computation complexity. Furthermore, offspring trees may have worse fitness than the parent trees as new edges are introduced.

**Exhaustive Recombination Operator.** The Exhaustive heuristic's performance heavily depends on the similarity of both parent trees as otherwise an expensive search in the solution space will be performed. Especially for setups with random initial trees, recombining is far too expensive resulting in little to no improvement within the given time bounds. Regarding the Raidl instances, only for trees based on the MST or AM-C heuristic this recombination operator results in trees comparable to those found by other recombination operators, for non-Euclidean random instances this is true only for AM-C-based trees. Starting from MST initial solutions, setups on Raidl instances converge slower with increasing population size, whereas for random instances setups converge faster with increasing population size. E. g. for `Raidl.100` and population size 2, the average excess is 0.03 %, whereas for population size 8, it is 2.07 %. For `Rand.500.E` the excess decreases from 9.33 % to 7.44 %.

**Path Merging Recombination Operator.** The Path Merging operator improves a solution slowest but constantly in most setups, but eventually reaches good near-optimum solutions given enough time. Regarding population size, this recombination operator finds on average better results with smaller population sizes (2) than with larger populations (8) in all cases except for AM-C setups with Raidl or non-Euclidean random instances. E. g. for `Raidl.100` and starting from random trees, setups with population size 2 are 3.40 % above the optimum on average, but have an excess of 5.70 % for population size 8. For random Euclidean instances, PM is the best performing operator next to RPL.

**Replacement Recombination Operator.** To evaluate the recombination operators above, experiments with a Replacement operator have been conducted. RPL's performance relies on the local search's performance only. As no diversification is introduced, early convergence can be expected when starting from poor initial solutions. E. g. for `Raidl.100` starting from random trees, the average solution excess is 3.86 %, 2.88 %, and 4.75 % (2, 4, and 8 individuals) using the replacement recombinator, whereas a more sophisticated recombination operator such as the Tree Building recombinator achieves solutions of excess 0.13 %, 0.12 %, and 0.04 %, respectively, on average. Still, the replacement operator is the best recombination setup for random instances using Euclidean distances.

**Global Trends.** For all instances where the AM-C heuristic performs best in constructing trees, all four recombination operators perform better with larger population size. E. g. for `Rand.300` and the PM operator, the average final excess is 1.14 % when using a population size of 2, whereas it is 0.23 % with a

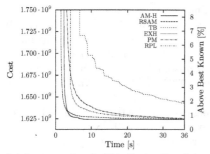
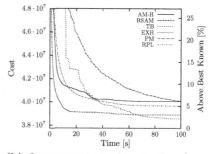
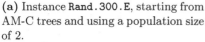

(a) Instance Rand.300.E, starting from AM-C trees and using a population size of 2.

(b) Instance Rand.500, starting from MST-based trees and using a population size of 4.

**Fig. 7.** Example plots visualizing the performance of different recombination setups in comparison to two iterated local search algorithms

population size of 8. Raidl and random instances with non-Euclidean distances show comparable behavior in similar setups, whereas random instances with Euclidean distances form their own group. E. g. TB is one of the best performing recombination operators for the former group, but performs worse than PM for the latter group. The memetic algorithm's convergence speed is slower compared to both ILS algorithms when applied to random instances with Euclidean distances (see Fig. 7a), but is able to surpass the ILS algorithms for non-Euclidean instances by using TB or PM, given enough time (see Fig. 7b).

## 5 Conclusions

We discussed a memetic algorithm for the OCST by evaluating aspects such as choice of initial solutions and recombination operator. For evaluation we used both real-world based instances (Raidl series) and randomly generated instances with and without Euclidean distances.

For instances with non-Euclidean (random) distances, our memetic algorithm outperforms the iterated local search algorithms AM-H and RSAM. Compared with other recent publications which mostly focus on pure evolutionary algorithms, our MA clearly outperforms these approaches. Our algorithm successfully processes instances with sizes not yet covered in previous publications.

Future work will focus on optimizing the presented algorithms and experimenting with larger instances. We are planning to develop a distributed memetic algorithm and to perform a search space analysis for non-Euclidean instances.

## References

1. Hu, T.C.: Optimum Communication Spanning Trees. SIAM Journal of Computing 3(3), 188–195 (1974)
2. Garey, M.R., Johnson, D.S.: Computers and Intractability: A Guide to the Theory of NP-Completeness. W.H. Freeman, San Francisco, CA, USA (1979)

3. Wu, B.Y., Chao, K.M., Tang, C.Y.: Approximation algorithms for some optimum communication spanning tree problems. Discr. Appl. Math. 102(3), 245–266 (2000)
4. Ahuja, R.K., Murty, V.V.S.: Exact and Heuristic Algorithms for the Optimum Communication Spanning Tree Problem. Transp. Sci. 21(3), 163–170 (1987)
5. Palmer, C.C., Kershenbaum, A.: Two Algorithms for Finding Optimal Communication Spanning Trees. Technical Report RC 19394, IBM T. J. Watson Research Center, Yorktown Heights, NY, USA (1994)
6. Palmer, C.C., Kershenbaum, A.: Representing trees in genetic algorithms. In: Proceedings of the First IEEE Conf. Evo. Comp., vol. 1, pp. 379–384 (1994)
7. Soak, S.M.: A New Evolutionary Approach for the Optimal Communication Spanning Tree Problem. E89–A(10), 2882–2893 (2006)
8. Bäck, T.: Evolutionary Algorithms in Theory and Practice. Oxford University Press, New York (1996)
9. Hoos, H.H., Stützle, T.: Stochastic Local Search: Foundations and Applications. The Morgan Kaufmann Series in Artificial Intelligence. Morgan Kaufmann, San Francisco (2004)
10. Merz, P.: Memetic Algorithms for Combinatorial Optimization Problems: Fitness Landscapes and Effective Search Strategies. PhD thesis, University of Siegen, Germany, Siegen, Germany (2000)
11. Rothlauf, F., Gerstacker, J., Heinzl, A.: On the Optimal Communication Spanning Tree Problem. Working Papers in Information Systems 10/2003, University of Mannheim, Germany (2003)
12. Fischer, T.: Improved Local Search for Large Optimum Communication Spanning Tree Problems. In: MIC 2007. 7th Metaheuristics International Conference (2007)
13. Rothlauf, F., Heinzl, A.: Developing efficient metaheuristics for communication network problems by using problem-specific knowledge. Technical Report 9/2004, University of Mannheim, Germany (2004)

# Hybrid Numerical Optimization for Combinatorial Network Problems

Markus Chimani, Maria Kandyba[*,**], and Mike Preuss[***]

Dortmund University, 44221 Dortmund, Germany
maria.kandyba@cs.uni-dortmund.de
http://ls11-www.cs.uni-dortmund.de

**Abstract.** We discuss a general approach to hybridize traditional construction heuristics for combinatorial optimization problems with numerical based evolutionary algorithms. Therefore, we show how to augment a construction heuristic with real-valued parameters, called *control values*. An evolutionary algorithm for numerical optimization uses this enhanced heuristic to find assignments for these control values, which in turn enable the latter to find high quality solutions for the original combinatorial problem. Additionally to the actual optimization task, we thereby experimentally analyze the heuristic's substeps.

Furthermore, after finding a good assignment for a specific instance set, we can use it for similar yet different problem instances, without the need of an additional time-consuming run of the evolutionary algorithm. This concept is of particular interest in the context of computing efficient bounds within Branch-and-Cut algorithms. We apply our approach to a real-world problem in network optimization, and present a study on its effectiveness.

## 1 Introduction

We consider a hybrid approach to use numerical evolutionary algorithms for solving NP-complete combinatorial network design problems. Most traditional hybridization approaches consist of developing combinatorial representations of solutions for use in evolutionary algorithms, or of developing meta-search heuristics which use traditional heuristics as subroutines. This results in algorithms which have, in general, a much longer running time than traditional combinatorial construction heuristics.

We divert from these approaches by taking an existing construction heuristic and augmenting it with *control values* (CVs). These are simple mostly non-discrete values, which allow us to modify the behaviour of the algorithm and

---

[*] Supported by the German Research Foundation (DFG) through the Collaborative Research Center "Computational Intelligence" (SFB 531).
[**] Corresponding author.
[***] Supported by the DFG project grant no. 252441, "Mehrkriterielle Struktur- und Parameteroptimierung verfahrenstechnischer Prozesse mit evolutionären Algorithmen am Beispiel gewinnorientierter unscharfer destillativer Trennprozesse".

T. Bartz-Beielstein et al. (Eds.): HM 2007, LNCS 4771, pp. 185–200, 2007.

thereby change its outcome: e.g., we may have a CV specifying a degree of sortedness when performing some heuristic step on a series of items; we may have a real-valued CV, as a balancing parameter to compute a weighted sum of two different optimization subcriteria, etc.

In general, we will not know in advance which settings for these CVs are in fact beneficial. We optimize the performance of the heuristic on a specific problem instance or instance set with a general-purpose optimization algorithm, namely an *evolutionary algorithm* (EA) designed to operate on numerical values. We find near-optimal settings $\chi^*$ for these CVs, using the augmented heuristic to compute the objective value for any specific CV instance $\chi$. Whereas this may appear as 'yet another EA application' on first sight, it is more than that, because

- the EA actively modifies the behaviour of the heuristic and identifying viable CV settings may not be reasonably possible without the composition with an optimization algorithm, and
- the performance feedback allows us to obtain new insight into the effectiveness of specific steps of the heuristic, which may lead to further ideas for improving the underlying algorithm.

In this paper we analyze this novel approach, by concentrating on a single application as a test case for the suggested high-level relay hybridization [7]. Nonetheless, we believe that this scheme may be successfully utilized for virtually every construction heuristic that contains components which have been added on the basis of ad-hoc or intuitive decisions. Our approach is particularly interesting in the realm of Branch-and-Cut algorithms.

Our test case is the real-world minimization problem [8] which arises when one wants to extend an already existing, city-wide fiber-optics or telecommunications network. This problem is known as the 2-root-connected prize-collecting Steiner network problem (2RPCSN) and has been studied, e.g., in [3, 8, 9]. In these papers, different integer linear programming approaches are proposed, which solve this problem to provable optimality, albeit in exponential worst-case time. These schemes use Branch-and-Cut techniques and employ heuristic algorithms as subroutines to generate upper bounds. The currently most successful such heuristic is described in [3]. The heuristic plays three distinct crucial roles for solving 2RPCSN in practice:

- Due to the NP-completeness of the problem, and the therefore exponential running times, computing a provable optimal solution is only feasible for small to medium sized instances. For large instances, one has to resort to heuristic approaches.
- ILP-based approaches benefit from good initial solutions. They often allow faster computation of LP relaxations, and can be used to obtain initial variable sets in column-generation based approaches, i.e., when certain variables of a linear programm are only added later if required.
- Branch-and-Cut approaches use upper bounds to cut off subtrees in their Branch-and-Bound trees. These upper bounds are computed using heuristics which can deal with the fact that certain variables are fixed at some

Branch-and-Bound node. Thereby, sophisticated bounding and efficient heuristics allow to solve larger instances and therefore increase the applicability of optimal algorithms.

Based on these demands, we can differentiate between different use-cases for our hybridization:

**Online:** The heuristic consists of running the evolutionary algorithm and outputs the best solution found. The CV optimization is thereby a vehicle to steer the search for different CV instances. The running time is much slower than using only the combinatorial construction scheme, but the expected quality of the solutions improves.

**Offline:** Alternatively, we can run the CV optimization on a set of test problem instances. The obtained CV instance $\chi$ can then be used to solve new problem instances. The selection of $\chi$ is therefore an educated guess which can be expected to be suitable for instances similar to the test problem instances. The major advantage is that the resulting running time is virtually the same as for the original construction heuristic.

**Mixed:** In the realm of Branch-and-Cut approaches, we can blend these two approaches in a very natural way: we can use the online variant to generate both a good initial solution for the problem instance, and also obtain a CV instance $\chi^*$ which is tuned specifically to this instance. During the time-critical steps within the branching strategy, we can use the offline variant with $\chi^*$ to obtain a good and fast bounding heuristic. Note that we do not have to restrict ourselves to a single $\chi^*$, but we can use, e.g., multiple CV instances from the last population of the evolutionary algorithm.

We try to answer the specific questions summarized in Section 2. Afterwards we decribe our general hybridization approach and the evolutionary algorithm in Section 3, before centering on our specific construction heuristic and its control values in Section 4. Section 5 and Section 6 focus on the conducted experiments and their analysis.

## 2 Aims

The presented hybridization for the above combinatorial network problem is ment as a proof-of-concept. Therefore our aims are centered on showing its success. Although we cannot guarantee the usefulness of this hybridization technique for completely different problems, we do suggest to try the approach if one can think of a way to make a specific heuristic configurable.

Before we can legitimatly consider one of the three outlined uses of our approach, we have to investigate if the hybridization is beneficial at all. Hence, the concrete aims pursued in this work are:

1. We want to determine if our EA is able to identify CV instances for the construction heuristic which improve its performance significantly both on single instances and on sets of similar instances. We thus ask if the heuristic can be automatically adapted to considered problems.

2. If Aim 1 can be positively answered, we want to identify where this adaptability stems from, i.e., which of the heuristic's steps contribute most to an improved performance, and which CVs can be safely set to default values.

As we expect that the heuristic can benefit from specific properties of certain problem instances, the detection of such properties is a side goal. Furthermore, the applied EA variant itself shall be assessed, to obtain a first impression concerning its usefulness for the given task. In the light of a unified EA approach [4], this resembles the question for a suitable parametrization of the EA.

## 3   Hybridization

In general, a non-trivial construction heuristic for an NP-hard problem consists of multiple steps. As the heuristic cannot guarantee to find an optimal solution, there may be multiple orthogonal reasons for non-optimality, e.g.:

- A single step may be not solvable to optimality in polynomial time, thus requiring a sub-heuristic.
- The optimization goal of some step $s$ may only be an educated guess in the context of the complete algorithm, i.e., the optimal solution for the step $s$ may in fact be subobtimal for the subsequent steps.
- Assume there are multiple steps, whereby their order is crucial for the outcome of the whole algorithm. The optimal order may be a priori unknown.

The idea is to identify the parts of the heuristic where some step is not based on theoretically provable facts, but on ad-hoc or statistical decisions. The algorithm is then modified such that these decisions can be controled via control values. We present some general cases of how to introduce CVs into construction heuristics. Our test application described in Section 4 will introduce a CV for most of the types below. The following list is of course by no means complete, but the list's items are meant as examples of a general design pattern.

**Selection Decisions:** When an algorithm has to choose between two possibilities, or select a subset of elements, we can introduce a simple CV $c \in [0,1] \subset \mathbb{R}$. Introducing randomization, we can interpret $c$ as the probability of choosing one out of two possibilities, or as the probability for an element to belong to the chosen subset. If the base set $S$ for the latter problem is ordered, we can also interpret $c$ directly as the selection ratio, without any randomization: we simply select the first $[c \cdot |S|]$ (rounded) elements.

**Balancing Decisions:** In many applications, there are subproblems which have to optimize multiple criteria at once and where it is unclear which of the different measures is most important for the overall optimization goal. We can introduce a CV $c \in [0,1] \subset \mathbb{R}$ to balance two such measures $x$ and $y$ and obtain a balanced value, e.g., by $z := c \cdot x + (1-c) \cdot y$. Note that, if applicable, the CV $c$ has not be used directly in a linear balancing computation, but we can use any suitable balancing function. Another balancing problem can occur when the input is partitioned in sets of different types, e.g., weights on nodes and edges of a graph. It might not be clear how much influence each set has for the overall optimization goal. This is, e.g., the case in our test application and is resolved via the *balance* CVs, see Section 4.

**Ordering Decisions:** Assume an algorithm where there is a sequence of steps and we do not know beforehand, which order will lead to the best solution. Traditional algorithms may choose some arbitrary order, often being the natural order arising from the order in which the input was given; other algorithms may randomize the order on purpose. Often, one has a certain guess, which order may be sensible, and sorts the steps accordingly. We can introduce a CV $c \in [0,1] \subset \mathbb{R}$ to steer this behaviour in two different ways:

*Bi-ordered.* Let there be two competing orderings $o_1$ and $o_2$ under consideration. Interpreting $c$ as the probability for selecting $o_1$, we can randomly choose the first, yet unselected element from one of these orderings. We can simulate a gradual randomization, by choosing $o_2$ as a random ordering.

*Random Range.* When mixing a potentially sensible ordering $o_1$ with a random ordering, we can interprete $c$ as a ratio of elements. Let $n$ be the number of ordered elements. For each step, we choose randomly from the first $m = \lceil c \cdot (n-1) + 1 \rceil$ elements induced by $o_1$. While $c = 0$ corresponds to the ordering $o_1$, $c = 1$ resembles a purely random order.

**Treshold Decisions:** Sometimes a heuristic step requires the identification of certain critical values, patterns, graph structures, etc. A CV can be used as a treshold to determine when a search pattern is considered to be identified. E.g., a simple treshold decision is the classification of fractional values into two groups of small and large values, respectively. One has to be careful with CVs in the context of identifying termination criteria of loops, etc., as a CV is purely for optimizing the solution quality, and not for optimizing the computation time: a CV may be suitable if there is no obvious connection between the number of iterations and the overall optimization goal. But if a CV would purely decide how often a loop, which iteratively improves the final solution, is run, a larger iteration count would always be superior to an earlier termination.

**Discrete Decisions:** There can be situations when augmenting a specific heuristic step with numerical values seems not reasonable. There may be intricately discrete decisions to make. Our approach is stable enough to allow integer CVs, although it comes with the usual problems known from optimizing integer parameters via an evolutionary algorithm, see below.

**Natural CVs:** Some construction heuristics already offer numerical parameters, e.g., for coarsening the input instance. We call these *natural CVs* and can directly use them within our optimization framework. Sometimes subproblems are solved via a PTAS, i.e., a polynomial approximation algorithm which approximates the solution within a choosable bound $\epsilon$. If it is not clear that an optimal solution for the subproblem will lead to an optimal solution for the whole problem, this $\epsilon$ can also be seen as a natural CV already in the original algorithm.

Note that the selection of the CVs not only changes the behaviour of the construction heuristic, but also directly influences the evolutionary algorithm and the applicability of the overall algorithm, due to time constraints: if we have many CVs, the EA will require more generations to find a near-optimal CV instance. If many CVs introduce statistical noise due to randomization techniques, the evolutionary algorithm will require more calls to the augmented heuristic to assess the quality of a single CV instance. Hence, we suggest to drop a CV $c$ when experiments show a clearly winning setting for $c$ or when it becomes clear that $c$ has no measurable influence at all.

### 3.1   Why Using Evolutionary Algorithms for CV Adjustment?

EAs are known as general-purpose direct optimization algorithms which only require a quality criterion (*fitness*) but no other additional data, e.g., on the gradient. They provide generic search operators for virtually every canonic representation, e.g., boolean, ordinal discrete, nominal discrete, and real-valued variables. As they are able to deal with mixed representations, they allow a 'natural' problem formulation and are therefore easy to apply. Emmerich et al. [5] give an example for a mixed integer problem and summarize the search operators for the basic variable types. Population-based EAs are by design a compromise between global and local search. If no good starting point is known, an EA usually starts with a small random sample and learns an internal model from the feedback obtained by subsequent search steps.

However, there are cases where a naïve EA does not perform well, namely on approximately random functions, and on simple unimodal functions. In the first case, there is no structure that could be learned, and in the second case, more specialized algorithms, e.g., quasi-Newton methods, are much faster. We assume that the CV optimization problem is neither of them and thus EAs are applicable. Furthermore, some of the previously described CV types introduce non-determinism into the target function evaluation, which can be seen as 'noise' from the viewpoint of the optimization algorithm. Recent investigations on noisy functions report that EAs still work reasonably well under such conditions [1].

Summarizing, it makes sense to apply an EA, not only because it can be easily done, but also since probably stronger analytical methods are unavailable as virtually nothing is known about the CV-induced search space topology. The EA shall perform as least as good as random search; if the EA is able to detect some exploitable structure, it will perform even better.

In our test application, we establish the adaptation of the CVs via a naïve $(\mu, \lambda)$ *evolution strategy* (ES) as described in [2]. According to the CV list given in Section 4, we have six real-valued and two nominal discrete optimization variables and thus a mixed genome. While this does not pose a problem for recombination which is performed as a 2-point crossover, the mutation operator requires special attention: we employ a commonly used normal distributed mutation for the real-valued variables and conditioned random selection for the nominal discrete variables. For the random selection, a mutation probability is used instead of the mutation strength, and if a change is determined, the new value is chosen randomly from all other allowed values but the current one.

Mutation step sizes and mutation probabilities, respectively, are controlled by self-adaptation, thereby learning strategy parameters from successful steps. As the domain sizes of the six real-valued variables are very similar—either 1 or 2, see below—we resort to a single strategy parameter, the mutation strength $\sigma$. It is varied by means of a learning rate $\tau$, which also applies to the second strategy parameter $\sigma_{dis}$, that resembles the mutation rate for the two nominal discrete variables. The chosen values for population size, offspring number, initial step sizes, minimum and maximum step sizes, and learning rate are given below. The last four have also been used for $\sigma_{dis}$. Currently, they have not been verified systematically as this would require too much computation power; instead, they have been tested against some representatative alternative parameter sets.

| $\mu$ | $\lambda$ | $\sigma_{\text{init}}$ | $\sigma_{\min}$ | $\sigma_{\max}$ | $\tau$ |
|-------|-----------|------------------------|-----------------|-----------------|--------|
| 15 | 100 | 0.3 | 0.001 | 0.5 | 0.2 |

## 4    The Network Problem and a Corresponding Heuristic

Formally, we can describe the 2-root-connected Prize-Collecting Steiner Network Problem (2RPCSN) as follows: We are given an undirected graph $G = (V, E)$, a root node $r \in V$, a set of customer nodes $C = C_1 \dot\cup C_2 \subset V$, a prize function $p : C \to \mathbb{R}^+$ for the customers, and a cost function $c : E \to \mathbb{R}^+$ on the edges. We ask for a subgraph $N = (V_N, E_N)$ of $G$ with $r \in V_N$ which minimizes $\sum_{e \in E_N} c(e) - \sum_{p \in C \cap V_N} p(v)$ and satisfies the following connectivity property: for every node $v \in C_k \cap V_N$, $k \in \{1, 2\}$, $N$ contains at least $k$ node-disjoint paths connecting $v$ to $r$. Informally, this means that we look for the most cost-efficient network, where we can choose which customers we want to connect, based on the connection costs $c$ and their estimated profits $p$. For certain customers $C_2$ we can only descide whether we want to connect them via two disjoint connection paths, or not at all.

**Overview.** We sketch the heuristic presented in [3], and explain its modifications in order to support certain control values: The heuristic starts by choosing subsets $C_i^* \subseteq C_i$ ($i = 1, 2$) of customers to be included in the solution. Based on this customer set $C^* := C_1^* \cup C_2^*$ we heuristically compute a minimal Steiner tree $T$ as a basis for the next steps. Note that $T$ resembles a feasible solution if $C_2^* = \emptyset$. We then extend this tree by adding additional paths for the $C_2^*$ customers, and obtain a feasible solution $S$. Finally, we perform some postprocessing to shrink $S$ by removing certain nodes and edges without losing feasiblity.

**Choose Customers and Construct $T$.** The heuristic is especially suited for the use within a Branch-and-Bound algorithm. Thereby intermediate fractional solutions are used to infer suitable sets $C_1^*$ and $C_2^*$. In [3], the setting $C_1^* = C_1$ and $C_2^* = C_2$ was suggested for the use as a constructive start heuristic. We introduce our first two real-valued control values:

$$choose_1, choose_2 \in [0,1] \subset \mathbb{R} \tag{1}$$

The CV $choose_k$, $k \in \{1,2\}$, determines the fraction of $C_k$ customers, which is chosen for $C_k^*$. I.e., having an ordered set of $C_k$ customers, we choose the first $[choose_k \cdot |C_k|]$ vertices as the set $C_k^*$. Of course the ordering of $C_k$ is crucial. This leads to two additional integer CVs. Note that the differences between real-valued and integer CVs are quite interesting, as we will see in Section 6.

$$sort_1, sort_2 \in \{0,1,2,3\} \subset \mathbb{N} \tag{2}$$

The interpretation of the image set of $sort_k$, $k \in \{1,2\}$, is as follows:

**0:** $C_k$ is permutated randomly.
**1:** $C_k$ is sorted by decreasing profit $p(v)$ of $v \in C_k$.
**2:** $C_k$ is sorted by increasing cost $z(r \to v)$ of the shortest path from $r$ to $v \in C_k$.
**3:** $C_k$ is sorted by decreasing efficiency, which is defined as the ratio $\frac{p(v)}{z(r \to v)}$.

After choosing $C^*$, we apply the Minimum Steiner Tree heuristic by Mehlhorn [6] to compute the tree $T$. This algorithm requires the shortest paths between all pairs of nodes of $C^* \cup \{r\}$. In [3], these shortest paths are computed using only the edge costs $c$. This approach does not take into account that a path might include a customer node which offers some profit and therefore renders the costly path cheap or even profitable.

Based on the observation that an inner node of a path is incident to exactly two edges of the path we would like to use the modified edge costs $c'(u,w) := c(u,w) - \frac{1}{2}(p(u) + p(w))$. However, $c'$ cannot be easily used for shortest paths computations, as its values may be negative and thus induce negative cycles. On the other hand, using $c'' := \max\{c', 0\}$ as a cost function is problematic as well: all edges incident to a profitable vertex will have the same cost of 0, and the algorithm is likely to choose a random edge instead of making a well-grounded descision. We therefore use the cost function $c^*(u,w) := \max\{c(u,w) - \alpha\frac{1}{2}(p(u) + p(w)), 0\}$, using the real-valued $\alpha \in [0,1]$ to balance the influence of the customer profits. As initial experiments showed, $\alpha = 0.5$ and $\alpha = 1$ lead to very similar cost functions, since in both cases edge costs are often truncated to 0. Hence we do not choose $\alpha$ as a CV directly, but use the control value $balance_T$ which is then transformed into a suitable value for $\alpha$:

$$balance_T \in [0,1] \subset \mathbb{R} \tag{3}$$

To make up for the aformentioned skew in the $\alpha$ values, we use $\alpha = balance_T^3$. Note that we use the resulting cost function not only within the minimum Steiner tree heuristic, but also for the sorting of $C_k$ if $sort_k \geq 2$.

**Assure 2-Connectivity.** The idea is to iteratively extend $T$ by adding shortest $(v \to r)$-paths for the customer vertices $v \in C_2^*$ for which the 2-connectivity requirement is not satisfied yet. Let $v$ be such a customer and let $P_v$ the inner nodes of the unique $(r \to v)$-path in the original $T$. To find a disjoint second path between $r$ to $v$, we look at the graph $G \setminus P_v$, i.e., the graph obtained by removing all inner nodes of the original tree path, and all their incident edges. We set the costs of the edges which are already in the solution to 0, and use the cost function $c^*$ for the other edges. We then apply Dijkstra's shortest path algorithm to identify a path between $v$ and $r$, and add the resulting path to our solution. This step uses its own control value

$$balance_S \in [0,1] \subset \mathbb{R} \tag{4}$$

to skew the cost function $c^*$, analogously to $balance_T$. This augmentation is performed for all nodes $v \in C_2^*$ which do not contain any further $C_2$ customers in their subtrees of $T$; let $L$ be the list of these $C_2^*$ nodes. Note that 2-connecting $v$ results in proper 2-connectedness for all nodes $w \in P_v$.

The quality of the resulting subgraph $S$ can depend on the order of the nodes in $L$. An intuitive idea, as proposed in [3], is to sort these nodes $v$ in decreasing order by the number of $C_2$ customers in $P_v$. If this number is identical for different nodes, we sort those by decreasing overall number of elements in $P_v$. However, experiments show that choosing another order can change and improve the solution quality. Hence, we introduce the real-valued control value

$$order^+ \in [-1,1] \subset \mathbb{R}. \tag{5}$$

Let $L_r$ and $L_o$ be copies of the list $L$. While $L_r$ is a random permutation, the list $L_o$ is sorted as described above. If $order^+$ is negative, we will reverse $L_o$. The absolute value of $order^+$ indicates the probability of taking the next node from $L_o$; otherwise the next node from $L_r$ is chosen.

**Shrinking.** In general, the subgraph $S$ obtained by the previous steps can be further optimized by removing some nodes and edges from $S$ without losing feasiblity.

We know that due to the construction, $S$ consists of one or more non-trivial 2-connected components, which have only the root node $r$ in common. All other components of the graph form trees, which are attached to some 2-connected component. Having this decomposition in mind we can optimize $S$ in two steps:

As described in [10], the rooted price-collecting Steiner tree problem can be solved to optimality in linear time, when applied to trees by dynamic programming. We use this algorithm to optimize all attached trees, using the attachment node as its root. For the next step, these root nodes are considered to be $C_1$ customers with corresponding prizes. These optimizations are optimal and independent of each other, and hence there is no need for any CVs.

In the final step we try to optimize each non-trivial 2-connected component $B$ of $S$. Such components may contain redundant edges, i.e., their removal does not break the feasibility of the solution. For every $B$ we compute its *core graph* $\tilde{B}$. Thereby, every chain of edges only containing nodes $v \in V \setminus (C_2 \cup \{r\})$ is replaced by a single edge. We then successively consider the edges $e$ of $\tilde{B}$: it can be removed from $\tilde{B}$ if all connectivity requirements are still satisfied for $\tilde{B} - e$. In this case, the corresponding path $P_e$ will be broken apart and partially removed. The optimization within the path $P_e$ can be done optimally and efficiently.

Similar to the steps before, the order in which the core edges are considered and removed from $\tilde{B}$ may be crucial for the solution quality. An intuitive sorting criterion is to use the weight of the core edge, i.e., the sum of the edges in $P_e$ minus the sum of the profits of inner nodes of $P_e$. We parameterize this order by a real-valued control value:

$$order^- \in [-1, 1] \subset \mathbb{R} \qquad (6)$$

Again, we have a randomly permutated and a sorted list of core edges. A positive $order^-$ leads to a decreasing order, a negative value leads to an increasing order. The absolute value of $order^-$ gives the probability of choosing the next core edge from the ordered list, instead of from the randomly permutated list.

Overall, we can give the CV instance which resembles the behaviour of the original heuristic. Technically, the original heuristic does not use any sorting to obtain $C^*$: since the *choose* CVs are 1, these parameters have no influence.

| $choose_1$ | $choose_2$ | $sort_1$ | $sort_2$ | $balance_T$ | $balance_S$ | $order^+$ | $order^-$ |
|---|---|---|---|---|---|---|---|
| 1 | 1 | any | any | 0 | 0 | 1 | 1 |

## 5   Experimental Assessment of Performance Improvement

We test our hybrid algorithm on three different test sets, which were used and described in [3]. The groups $K$ and $P$ consist of graphs each containing 400 nodes and a high percentage of customer nodes, allowing us to better analyze the influence of the *choose* CVs. In particular, the set $K$ consists of random geometric instances which were designed to have a structure similar to street maps. All of these instances could be solved to optimality with the Branch-and-Cut approach presented in [3]. These instances are chosen in order to evaluate the absolute solution quality achieved by our algorithm. Additionally, we test against the instance set $ClgM^+$. These instances are real-world graphs based on a street map of Cologne. The graphs have 1757 nodes and relatively small number of customers (up to 15–20 $C_1$ and $C_2$ customers, respectively). As these instances are hard for the ILP-approach—only one instance could be solved to optimality within 2 hours—we are in particular interested in improving the heuristic solutions and in analyzing our approach for its applicability within the mixed scenario (cf. Section 1).

The following two experiments investigate the question raised as Aim 1 in Section 2, i.e., does the hybrid approach improve the solution quality when applied to a single instance or to a group of instances.

### 5.1   Experiment 1: Does the Hybrid Approach Improve the Performance over the Default Heuristic on Single Problem Instances?

**Pre-experimental Planning.** By initial experiments, we found out that the run-length of the EA should be limited to at most 10000 evaluations for two reasons: firstly, the absolute running times, especially for small problem instances, are getting too large, and secondly, the optimization typically stagnates after ca. 5000–6000 evaluations. We also tried different numbers of repeats for evaluating a single CV instance, to smooth the noise introduced by the randomization. It turned out that 5–10 repeats seem to be a good compromise between speed and fitness value stability; we perform 10 repeats for the following experiment.

**Setup.** To test the performance of our hybrid algorithm on single problem instances, we perform 10 repeats for every problem instance and compute the best and the average objective values over these 10 independent solutions.

**Task.** We are interested in the improvement of the solutions—both the best and the average value—compared to the default heuristic, as described in Section 4. In order to state that the configured hybrid heuristic performs better than the default heuristic, we require that a Wilcoxon rank sum test is significant at the 5% level at least for the best of the 10 resulting CV instances for each problem instance, over 100 validation runs of the heuristic.

We are also interested in the solutions' quality with respect to the known optima [3]. We try to identify successful and unsucessful steps within the heuristic, which is useful when applying the heuristic to larger problems. Although we do not know the optima for any but one instance of the $ClgM^+$ group, we have lower and upper bounds stemming from two hours of Branch-and-Cut computation: they help to estimate the quality of our solutions in these cases.

**Results & Observations.** See Figure 1 for vizualizations of our experimental results. For various instances, the default heuristic is not able to find any feasible solution other then the trivial *root-solution*, i.e., no customers are selected. Such root-solutions are 115–689% away from the corresponding optimum. In contrast, our hybrid algorithm always computes at least one non-trivial feasible solution for all test instances. For the $K$ and $P$ instances, these solutions are 4–8% away from the optimum, on average; the gap is 11–15% for the non-trivial solutions of the default heuristic. The situation is similar for $ClgM^+$, where the default heuristic can only find the root-solution for two instances. Thereby the improvement by the hybrid algorithm is about 90%. For the other $ClgM^+$ instances the improvement is between 5 and 18%.

**Discussion.** The obtained results show that our approach reliably leads to good approximations, even when the default heuristic fails completely. By tuning certain parts it is therefore possible to obtain a better configured heuristic. The applied statistical tests confirm a significant improvement, as all p-values except for K400-8 are below $10^{-5}$. In this special case, a significant improvement is still detected, but slightly below the required level (p-value 0.145). We also see that the difference between the best and the average performance of 10 adapted heuristics is rather small. Hence running the EA one or few times seems to be sufficient to exploit most of the available performance potential.

**Additional Validation.** In order to verify that the EA performs consistently at least as well as the random search and often better, we applied both methods to

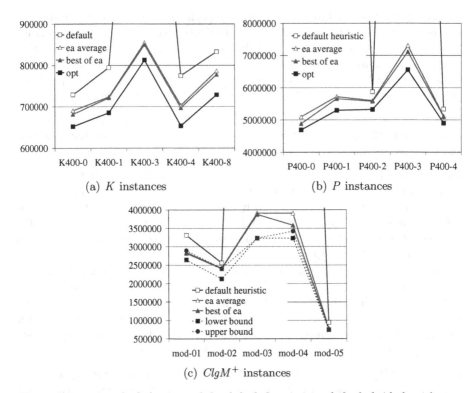

(a) $K$ instances

(b) $P$ instances

(c) $ClgM^+$ instances

**Fig. 1.** Comparing the behaviour of the default heuristic and the hybrid algorithm on the single test instances

three problem instances for which the default heuristic fails completely, namely K400-3, P400-3, and $ClgM^+$-03. The comparison of the mean best values of 20 repeats shows that indeed, the EA always achieves the same or a better fitness level. However, due to large standard deviations, Wilcoxon rank sum tests do not get significant at the 5% level (p-values 0.47, 0.91, and 0.08, respectively). Nonetheless, we can see a wide performance gap for the largest instance $ClgM^+$-03. This seems to be due to the fact that there is some potential for further improvement, whereby for the two smaller instances, both methods already operate near the optimum.

### 5.2   Experiment 2: Does the Hybrid Approach Lead to Improved Performance on Problem Instance Sets?

**Pre-experimental Planning.** Experiments show that 10 evaluations for each problem of a set would severely slow down the EA. As computing less generations seemed not sensibel, we resort to a 'weighted optimistic average' scheme that uses only 2 evaluations per instance and averages over the best of these two. This is necessary to avoid distorting effects of bad outliers. The average is weighted relative to the best fitness obtained on each problem during the computation of the first generation; we do this to avoid scaling effects stemming from different magnitudes of the instances' objective values.

**Task.** We accept the adapted heuristic as dominant to the default heuristic, if the resulting best CV instance leads at least to the same performance as the latter for every single problem instance of the test set. Additionally, we require the same for additional instances of the same type, which are used during the EA optimization. Overall, we require that statistical testing reveals a significant advantage on at least half of the tested instances per group.

**Setup.** We only consider the $K$ and $ClgM^+$ problem sets. During the validation phase of the best found CV instance, we add 2 previously unused $K$ instances and one $ClgM^+$ instance, respectively.

**Results & Observations.** The validation of the obtained best and average CV instances revealed that training with a full set instead of a single problem instance leads to clearly worse performance than reported in Experiment 1. The average CV instance leads to worse solutions than the default heuristic for several problem instances. The best CV instance however appears to be better than the default heuristic in most cases.

**Discussion.** The Wilcoxon rank sum test confirmes significant performance differences between the best obtained CV instance and the default heuristic, in all cases except two. The two critical problem instances are $ClgM^+$-4 with a p-value of 0.036, and $K400$-3 with a p-value of 1. Whereas in the first case, the attained advantage so small that it may stem from lucky sampling, we cannot report any improvement at all for the second case. However, for all other problem instances, we obtain significant differences with $p < 10^{-5}$, i.e., the CV instance leads to improved performance even on the instances that have not been used for training. We can therefore state that learning a suitable CV instance for a set of problem instances is successful, although the performance gain is not as large as for single problem instances.

# 6    Experimental Analysis of Adaptability

This section investigates the changes induced to the heuristic by automatically adapting it to the given instance or instance set. This corresponds to Aim 2 formulated in Section 2. The results documented in Section 5 clearly suggest that for a given instance, certain CV settings may be especially useful. Hence, for each instance we analyze the CV instances computed by the hybrid algorithm. Figure 2 shows the results for some representative problem instances.

Before analyzing each single CV, we may observe that there are certain evident dependencies between the CVs. E.g., if a *choose* CV is set to 0 or 1, the corresponding *sort* CV does not have any importance. Analogously, $choose_2 = 0$ makes both *order* CVs insignificant. We see such an effect in Figure 2(a). We know that it is difficult for our EA to handle nominal discrete CVs as they do

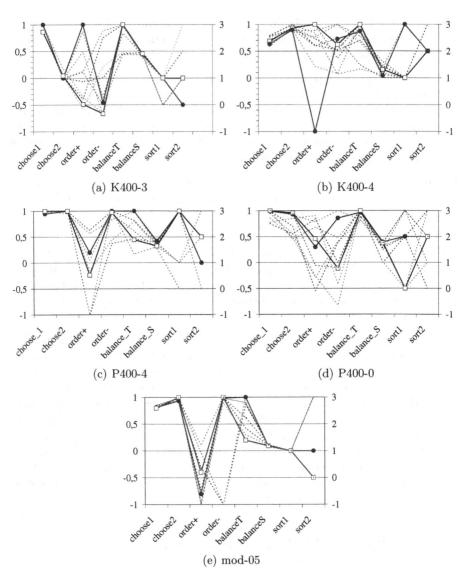

(a) K400-3

(b) K400-4

(c) P400-4

(d) P400-0

(e) mod-05

**Fig. 2.** CV instances for representative problem instances. The two solid lines identify the CV instances for the best solution (solid circles) and for the solution which has a value nearest to the average solution value (empty squares)

not provide any gradient information but simply have to be tried out. However, additional experiments show that their value in optimal CV instances is in fact meaningful: by changing a *sort* CV in a given CV instance, the solution quality usually decreases dramatically. For most instances where $choose_1 \neq 1$, we observe that the optimal CV instances have $sort_1$ set to either 1 or 3. We can deduce that sorting $C_1$ customers by their weight or by their efficiency is superior to random sorting or sorting by distance.

We can obtain much insight about the problem instance K400-3 by analyzing its CVs: $choose_2$ is very small, indicating that it is either not worthwhile to select any $C_2$ customer or it is not possible to connect them feasibly. The fact that the default heuristic cannot find any feasible solution suggests the latter.

The *balance* CVs are probably the most interesting control values, as we had no idea for a suitable setting, prior to the experiments. By interpreting their settings, we obtain some insight into the involved construction heuristic steps. These CVs specify how much the node profits should be considered for the shortest path computations. Most suprisingly, it turns out that there is a significant difference between $balance_S$ and $balance_T$. While the former is near to 1 for most problem instances, the latter usually is between 0 and 0.5. Although this fact may be quite surprising at first sight, there is a natural interpretation. Remember that a large *balance* value will result in paths which contain more profitable edges, but has the drawback that the edges incident to a profitable node become cost-wise indistinguishable. The CV $balance_T$ is used when computing shortest paths in the whole graph. The density of the node customers on a single path in the full graph is rather small, and choosing the perfect incident edges is not as important as the benefit of attaching a profitable node. This is not the case for $balance_S$: when using this CV we look for rather short and local connection path, whereby it is more important to choose optimal edges.

The diagrams also show that the *order* CVs do not clearly influence the solution quality. Although some problem instances indeed exhibit certain tendencies, these cannot be generalized. Therefore, we can set these CVs to default values, if we have no knowledge about the specific problem instance we are dealing with.

# 7    Conclusions

We introduced a general *CV hybridization* scheme, which allows to utilize numerical-based evolutionary algorithms for combinatorial problems. We showed that our hybridization approach works well for our test application, and it therefore seems reasonable to suggest to try this approach for different combinatorial problems and heuristics. We can also conclude that CV hybridization should focus on real-valued CVs, as nominal discrete CVs pose difficulties in general.

We assume that an analytical optimization of the parameters for the EA, i.e., population size, offspring number, etc., might lead to even further improvements.

# References

1. Arnold, D.V., Beyer, H.-G.: A comparison of evolution strategies with other direct search methods in the presence of noise. Computational Optimization and Applications 24(1), 135–159 (2003)
2. Beyer, H.-G., Schwefel, H.-P.: Evolution strategies—A comprehensive introduction. Natural Computing 1, 3–52 (2002)
3. Chimani, M., Kandyba, M., Mutzel, P.: A new ILP formulation for a 2-connected prize collecting steiner network problem. In: Proc. ESA 2007 (to appear, 2007)

# Author Index

# Lecture Notes in Computer Science

Sublibrary 1: Theoretical Computer Science and General Issues

For information about Vols. 1– 4488
please contact your bookseller or Springer

Vol. 4641: A.-M. Kermarrec, L. Bougé, T. Priol (Eds.), Euro-Par 2007 Parallel Processing. XXVII, 974 pages. 2007.

Vol. 4639: E. Csuhaj-Varjú, Z. Ésik (Eds.), Fundamentals of Computation Theory. XIV, 508 pages. 2007.

Vol. 4638: T. Stützle, M. Birattari, H. H. Hoos (Eds.), Engineering Stochastic Local Search Algorithms. X, 223 pages. 2007.

Vol. 4630: H.J. van den Herik, P. Ciancarini, J. Donkers (Eds.), Computers and Games. XII, 283 pages. 2007.

Vol. 4628: L.N. de Castro, F.J. Von Zuben, H. Knidel (Eds.), Artificial Immune Systems. XII, 438 pages. 2007.

Vol. 4627: M. Charikar, K. Jansen, O. Reingold, J.D.P. Rolim (Eds.), Approximation, Randomization, and Combinatorial Optimization. XII, 626 pages. 2007.

Vol. 4624: T. Mossakowski, U. Montanari, M. Haveraaen (Eds.), Algebra and Coalgebra in Computer Science. XI, 463 pages. 2007.

Vol. 4623: M. Collard (Ed.), Ontologies-based DataBases and Information Systems. X, 153 pages. 2007.

Vol. 4621: D. Wagner, R. Wattenhofer (Eds.), Algorithms for Sensor and Ad Hoc Networks. XIII, 415 pages. 2007.

Vol. 4619: F. Dehne, J.-R. Sack, N. Zeh (Eds.), Algorithms and Data Structures. XVI, 662 pages. 2007.

Vol. 4618: S.G. Akl, C.S. Calude, M.J. Dinneen, G. Rozenberg, H.T. Wareham (Eds.), Unconventional Computation. X, 243 pages. 2007.

Vol. 4616: A. Dress, Y. Xu, B. Zhu (Eds.), Combinatorial Optimization and Applications. XI, 390 pages. 2007.

Vol. 4614: B. Chen, M.S. Paterson, G. Zhang (Eds.), Combinatorics, Algorithms, Probabilistic and Experimental Methodologies. XII, 530 pages. 2007.

Vol. 4613: F.P. Preparata, Q. Fang (Eds.), Frontiers in Algorithmics. XI, 348 pages. 2007.

Vol. 4600: H. Comon-Lundh, C. Kirchner, H. Kirchner (Eds.), Rewriting, Computation and Proof. XVI, 273 pages. 2007.

Vol. 4599: S. Vassiliadis, M. Berekovic, T.D. Hämäläinen (Eds.), Embedded Computer Systems: Architectures, Modeling, and Simulation. XVIII, 466 pages. 2007.

Vol. 4598: G. Lin (Ed.), Computing and Combinatorics. XII, 570 pages. 2007.

Vol. 4596: L. Arge, C. Cachin, T. Jurdziński, A. Tarlecki (Eds.), Automata, Languages and Programming. XVII, 953 pages. 2007.

Vol. 4595: D. Bošnački, S. Edelkamp (Eds.), Model Checking Software. X, 285 pages. 2007.

Vol. 4590: W. Damm, H. Hermanns (Eds.), Computer Aided Verification. XV, 562 pages. 2007.

Vol. 4588: T. Harju, J. Karhumäki, A. Lepistö (Eds.), Developments in Language Theory. XI, 423 pages. 2007.

Vol. 4583: S.R. Della Rocca (Ed.), Typed Lambda Calculi and Applications. X, 397 pages. 2007.

Vol. 4580: B. Ma, K. Zhang (Eds.), Combinatorial Pattern Matching. XII, 366 pages. 2007.

Vol. 4576: D. Leivant, R. de Queiroz (Eds.), Logic, Language, Information and Computation. X, 363 pages. 2007.

Vol. 4547: C. Carlet, B. Sunar (Eds.), Arithmetic of Finite Fields. XI, 355 pages. 2007.

Vol. 4546: J. Kleijn, A. Yakovlev (Eds.), Petri Nets and Other Models of Concurrency – ICATPN 2007. XI, 515 pages. 2007.

Vol. 4545: H. Anai, K. Horimoto, T. Kutsia (Eds.), Algebraic Biology. XIII, 379 pages. 2007.

Vol. 4533: F. Baader (Ed.), Term Rewriting and Applications. XII, 419 pages. 2007.

Vol. 4528: J. Mira, J.R. Álvarez (Eds.), Nature Inspired Problem-Solving Methods in Knowledge Engineering, Part II. XXII, 650 pages. 2007.

Vol. 4527: J. Mira, J.R. Álvarez (Eds.), Bio-inspired Modeling of Cognitive Tasks, Part I. XXII, 630 pages. 2007.

Vol. 4525: C. Demetrescu (Ed.), Experimental Algorithms. XIII, 448 pages. 2007.

Vol. 4514: S.N. Artemov, A. Nerode (Eds.), Logical Foundations of Computer Science. XI, 513 pages. 2007.

Vol. 4513: M. Fischetti, D.P. Williamson (Eds.), Integer Programming and Combinatorial Optimization. IX, 500 pages. 2007.

Vol. 4510: P. Van Hentenryck, L.A. Wolsey (Eds.), Integration of AI and OR Techniques in Constraint Programming for Combinatorial Optimization Problems. X, 391 pages. 2007.

Vol. 4507: F. Sandoval, A.G. Prieto, J. Cabestany, M. Graña (Eds.), Computational and Ambient Intelligence. XXVI, 1167 pages. 2007.

Vol. 4502: T. Altenkirch, C. McBride (Eds.), Types for Proofs and Programs. VIII, 269 pages. 2007.

Vol. 4501: J. Marques-Silva, K.A. Sakallah (Eds.), Theory and Applications of Satisfiability Testing – SAT 2007. XI, 384 pages. 2007.

Vol. 4497: S.B. Cooper, B. Löwe, A. Sorbi (Eds.), Computation and Logic in the Real World. XVIII, 826 pages. 2007.

Vol. 4494: H. Jin, O.F. Rana, Y. Pan, V.K. Prasanna (Eds.), Algorithms and Architectures for Parallel Processing. XIV, 508 pages. 2007.

Vol. 4493: D. Liu, S. Fei, Z. Hou, H. Zhang, C. Sun (Eds.), Advances in Neural Networks – ISNN 2007, Part III. XXVI, 1215 pages. 2007.

Vol. 4492: D. Liu, S. Fei, Z. Hou, H. Zhang, C. Sun (Eds.), Advances in Neural Networks – ISNN 2007, Part II. XXVII, 1321 pages. 2007.

Vol. 4491: D. Liu, S. Fei, Z.-G. Hou, H. Zhang, C. Sun (Eds.), Advances in Neural Networks – ISNN 2007, Part I. LIV, 1365 pages. 2007.

Vol. 4490: Y. Shi, G.D. van Albada, J.J. Dongarra, P.M.A. Sloot (Eds.), Computational Science – ICCS 2007, Part IV. XXXVII, 1211 pages. 2007.

Vol. 4489: Y. Shi, G.D. van Albada, J.J. Dongarra, P.M.A. Sloot (Eds.), Computational Science – ICCS 2007, Part III. XXXVII, 1257 pages. 2007.